Canine Transcendence

The Scientific Impact of The Canidae Species

Michael Phife

Cover by Hyperlight Artwork

ISBN-13:
9780692050927

Dedicated to all the dogs and wolves that
have made an impact on my life

Contents

1. ORIGIN OF CANINES...1

2. DOMESTICATION...11

3. DEVELOPMENTAL STAGES............................29

4. FACING ENDANGERMENT 41

5. HYBRIDS...59

6. WILD DOGS 67

7. SYMBIOTIC RELATIONSHIPS 77

8. CANINE DISEASES & OUTBREAKS 89

9. FORENSIC SCIENCE 112

10. WOLF SANCTUARIES......................................126

ABOUT THE AUTHOR 138

INDEX 139

1
Origin of Canines

When talking about members of the canine family, there are several mammals that fit into this category. Most people think about wolves, dogs and coyotes as the main contributors to the family of canidae species. While wolves and dogs are the focal point of this book, there are several other species that I would like to give reference to in the regard of being their family members.

Foxes are relatives to the wolf and can be found worldwide in a multitude of habitats. In Africa and Eurasia we have the jackal, which has been depicted in ancient Egyptian culture, such as the god of Anubis. South America is home to the maned wolf, which isn't a true wolf but a very distinct canine member of a separate lineage. Also in South America and a few parts of Central America we have bush dogs, which are small in size and lack a lot of traits that other canines have. Australia is home to a species of feral dogs referred to as dingoes. Wild dogs can be found worldwide and are called a variety of names, such as the dhole from Asia, and the painted dog from Africa. Then in Eastern Asia and parts of Europe we have the raccoon dog, more commonly known as the tanuki in Japan.

All of these canines have a variation of characteristics with one another. They are all carnivores or scavenger-carnivores. Each member of the canidae family has an analogous anatomical structure with their ears, tails and muzzles all being very similar per species. However, there are also unique traits each species has which are separate from other genera of canines. Individual traits could have been altered through adaptation, by their location in the world, or through some type of evolutionary change in genetics over time. These adaptations also make it easy to compare and contrast dogs to their wolf relatives.

But before delving into the wolves and dogs of today, we need to look into the past. Where they originated from, and what has led to all of their genetic changes over time? It's obvious by looking at a canine's skeletal structure, especially the skull, that they are descendants from carnivorous species. This is easily proven by their teeth alone. Carnivores have very distinct incisors compared to herbivores. This should be a universal observation, but I'll contrast the differences for further clarity.

A carnivore's incisors and canine teeth are sharp and cone-like in shape. These are used to help puncture and kill their prey when they hunt. The premolars towards the back of the mouth are somewhat jagged or rigid, which helps break down meat to be swallowed and digested.

In contrast, most herbivore's teeth have a flattened or rectangular appearance. They are shaped in this manner because there is not a lot of grinding necessary for a primarily vegetation-concentrated diet. Omnivorous species will have a varying dental structure such as humans. Look at your own molars, and you will see a good representation of an herbivorous tooth set. Your canines would be more of what a carnivorous tooth set looks like.

Carnivorous mammals will also have a fairly pointed skull from the frontal lobe to the nasal cavity, almost triangular in shape for some species. This specific trait accommodates a heightened sense of smell, which aids in hunting. Although some herbivorous mammals will have a skull of similar frame, it becomes easier to distinguish between the two when looking at the skull itself rather than the exterior of the animal. Courses in zoological anatomy and physiology present this information in an anatomical way, making it much easier to contrast.

Now that the skull and dental structure is out of the way, we can get back into how canines have evolved into the species they are now. The wolf is the direct ancestor of all major canidae species. They have been around much longer than any other species of canine, roughly dating back one million years. With that taken into consideration, we can create a timeline based on how far wolf lineages date back and where they originated from specifically.

Roughly 62 million years ago there was a mammal that roamed the earth known as the Miacid. Their remains were initially found in North America and Europe, and believed to have migrated into Asia later on. This animal is what has been widely believed to be the origin of all known canines and felines. Yes, dogs and cats share a common ancestor, which is not surprising given how they have shared a lineage together throughout most eras of human history. Miacids were similar in size to that of large prehistoric rodents or the common dog of today. They were obligate carnivores, meaning they could not rely on vegetation as part of their diet.

About 20 million years later, the Miacids started

branching off into different species. The true descendants of Miacids today are animals known as genets and civets. For those of you who are unaware of what these animals are, genets resemble a cross between a cougar and a ferret, and civets look like a mix between a raccoon and a dog. These species are found predominately in Africa, but some are also found in southern Eurasia. Genets and civets may be the direct and least altered species to descend from Miacids, but the first canids arose from this prehistoric species as well.

I'd like to point out that the various species we see today, although descendants of Miacids, no longer share the same genetic family. This is not to cause confusion thinking that cats, civets and genets are related to canines. None of these animals are related genetically due to genetic drift.

Within these 20 million years of advancement, the first wolf emerged through a mutation of the Miacids genetics. Felines also branched off during this time period into what we know as the saber-toothed cat and then continued its evolutionary chain from there. From what we have been able to tell in regards to carbon dating and DNA, the first wolves had the ability to climb trees, much like felines. This was a trait that carried over from the Miacidae, as they were an arboreal carnivore.

The first wolves were very slender in build as well, some of the fossils resembling that of a large fox. There isn't much known about the original wolves, but we are able to draw conclusions based on their fossil records and skeletal structure. This early species of wolf lived for another 17 million years before it began evolving into various canid species, evolving to be closer in appearance to what we now know as a canine. These changes continued until one million years ago, which is when the gray wolf first emerged.

At some point, wolves migrated to today's North America by a land bridge and started increasing their population sizes. This migration is theorized to have happened 750,000 years ago, shortly after the modernized version of the gray wolf came to be. This movement to North America could have happened for several reasons. One reason being is that there could have been a change in climate, as the world started getting much colder due to transitioning into the Ice Age. This would also tie into a change in their diet, where smaller game was relatively abundant in the new world versus the old world.

The gray wolves of this time period have remained fairly similar in physical characteristics as the ones we see today. There have been some subtle differences, such as the

Armbruster's Wolf, which became extinct 300,000 years ago, and the Edward's Wolf, which also became extinct during this time period. The biggest differences were in skull composition and overall size. The Armbruster's Wolf had more of a ghastly cranium composition and was reasonably larger in size than other gray wolves during this timeframe. Whereas the Edward's Wolf had a slightly elongated jaw in comparison, but their build was nearly identical to other wolf species of this era.

It wasn't until much later that we saw a noticeable difference appear in the canine lineage, which began to separate gray wolves from the typical canine order. This shift in canine appearance was due to the introduction of the dire wolf. This is the wolf species that is renowned in folklore and legends all throughout North America. They first made their appearance roughly 125,000 years ago and were estimated to reach extinction around 16,000 to 4,000 years ago (with debates on fossil records). This was the largest canine to ever exist, standing over 2 feet (0.61 meters) tall and weighing an average of 150 pounds (68 kilograms). For comparison, the Irish wolfhound is equivalent to the dire wolf's height and close to its estimated weight.

Even though this animal was much larger, that does

not mean that it necessarily had advantages over the common gray wolf. For instance, their legs were stubbier than their counterparts, making their maneuverability and speed lesser in comparison. Their brain sizes were also significantly smaller, which meant that their hunting skills were not as keen as the standard wolf's. They were thought to be more of scavengers instead of hunters in this regard. However, the one advantage the dire wolf had over the gray wolf was a more powerful jaw. Their teeth were indicated to be so strong that they could crunch through bone, as observed by the indentions and chipping on their incisors and molars.

When it came to the species' longevity, the dire wolf simply could not compete against the gray wolf in terms of hunting, which ultimately led to their extinction. Its theorized that the dire wolf fed off the carcasses of animals that perished from the Ice Age. Once those animals completely died off, so too did this magnificent species. The gray wolf and its relative the coyote survived this wave of extinction, possibly by their cunning intellect and willingness to work in packs.

On that note, the coyote is thought to have existed for nearly 2.5 million years, around the time of the Ice Age. It is important to understand that the coyote is not a direct descendant from gray wolves. They are directly linked to gray wolves and other canids by an evolutionary clade; sharing a common ancestor which dates back several million years. Therefore, coyotes theoretically branched off from either the Miacid or from a very early wolf species.

As the Ice Age came to a close approximately 11,000 years ago, the wolf and coyote species were some of the few carnivorous mammals to make it out alive. While not every species of gray wolf survived this drastic climate change, such as the Beringian gray wolf, most members did. Gray wolves and coyotes began breeding and spreading well throughout North America shortly after the Ice Age came to an end. During this era, North America was connected to the rest of the world by a plot of land called Beringia. This was an area that linked today's Alaska with Russia, creating a full enclave around the Arctic.

Once the wolves migrated northward, they began heading from North America's arctic to the southern United States, Russia, and finally setting roots into Eurasia once again. This land bridge crossing is also how the Inuit people sojourned into North America nearly 18,000 years ago. It was also the first account of the Inuit's making contact with wolves.

Despite this migratory time period being the Inuit's

first encounter with wolves, it was not the first time humans had ever came into contact with a wolf. The first "modernized" humans have existed for roughly 50,000 years. The term modernized in this sense means people who were able to create a language, establish a civilization, work as a community and have records of their own history.

As canids have hit every major continent throughout time, humans were bound to have run into them during some point between the Ice Age and the Inuit's crossing. The dire wolf was reported by various tribes throughout history, with details of how massive they were in comparison to other wolves. So we know that humans must have encountered canines at some point early on during human development. This now gets us into the topic of canine interaction with man and the first cases of domestication.

2

Domestication

It is unclear when humans first encountered wolves, much less when we first started interacting with them as a means of survival. Wolves heavily rely on instinct and on their pack to ensure a continued survival. These are traits that both wolves and humans share together. We have survived as a species through population dependency and by initially using a hunter-gatherer mindset many years ago.

Wolves have survived extinction by working together as a pack. Every member of the pack has a role to play, and they all know their place the minute a pack is formed...or figure it out rather quickly. The alpha males and alpha females are the parents, generally seen as the leaders. They are the dominant members of the pack while all other wolves inside of the pack would be subordinate or submissive to them. The other members used to be classified as beta, delta and so on. These terms are considered controversial as wolf relationships are far more complex than initially realized. Smaller packs tend to keep these classifications, though, as they do follow a more precise hierarchy. The most submissive members in both large and small packs are still considered omegas, and are typically represented as the clowns or runts in the pack. This is not to say they are useless. On the contrary, their purpose is to help solve inner pack disputes and to break up fights that may occur.

Wolves in a hierarchy,
Wolf Creek Habitat

Alpha wolves are always the dominant and initial breeding pair in the pack, and their offspring become the latter. Occasionally, if a pack grows too large, a subordinate member will break off and form his or her own pack. If some of the wolves are not related then the pack can have multiple breeding pairs. Wolf packs with unrelated members almost never happen in the wild, though. This occurrence has only been seen in zoos and sanctuaries.

We can draw many similarities from pack roles within human hierarchy as well. The survival of our species has always been the greatest causation to work together and form social rank. In early human history we had leaders or chiefs, similar to today. They would oversee the workforce and survival of the collected group, much like an alpha wolf's role. Then the list trickles down from there with workers, hunters, gatherers, lookouts, caretakers, etc. Just as canines survived the Ice Age, we have survived several pandemics and epidemics that could have ended humanity as we know it. So why is all of this important to the study of wolves and how they came into contact with man?

One of the most important things that intelligent species look for in survival is companionship. They must be able to trust others to ensure that they have a longer lifespan. This is why you see animals such as deer, bison, birds, fish, and so forth in collected groups or herds. Their survival rate increases with more members in their party. Many of these species look for alternative ways to extend their survival with the help of companionship outside of their own species. This leads to something called a symbiotic relationship. An easy way to explain this is if one species were to help another species in exchange for something; quid pro quo. Quite a few animals do this in the wild, and this is how man and canine first found each other.

While there is no historical written report of when man first interacted with wolves, we can base theories or explanations off of fossil records. Among several archaeological findings, there have been human remains found alongside canine imprints that date back over 30,000 years ago. Most archaeologists and anthropologists believe that humans did encounter and work with wolves or some type of canine around this time period. Since this cannot be proven with absolute certainty, it is stationed as a disputed claim for now.

There is, however, more recent proof that ancient humans have indeed been found with canine remains. These findings date back nearly 20,000 years ago during the Ice Age in Eurasia. This helps strengthen the theory as to when humans first partnered up with wolves for hunting. What is interesting, though, is that the remains of these canines were not necessarily linked with the DNA of the wolves today. They actually had a plethora of canine genetics spliced into their remains. Their genetic composition was mixed with that of modern wolves, wild dogs and even coyote DNA. Therefore, it is presumed that the first acts of domestication could have been with a relative of wolves rather than with the true gray wolves that we see today.

Perhaps the first partnership between man and canine was out of necessity. The era of the Ice Age was a brutal time for all living creatures. Those who did not work as a team to overcome the harsh reality of the new world perished. Many animals died off during this era, and if a predator's food source dies off, then so too does the predator unless they are able to adapt. So rather than dying off with a vast number of mammalian species, wolves sought refuge in man, and vice versa.

There is one theory as to why wolves originally worked with humans, which both ties in and pieces together our initial relationship with them. Hunter-gatherer's during the time of the Ice Age would leave the remains of their food behind such as carcasses and bones as they ventured on. Wolves could have been drawn to the smell and followed humans from camp to camp in search of an easy meal. This could have led to an accidental partnership out of the wolves' necessity to survive rather than forming a companionship with the hunters.

Regardless of the reason, hunters soon realized that they could tame wolves and use them to help take down larger game. This bond continued and strengthened throughout the course of human development, and has been the first actual record of domestication with any animal. This is also what makes finding the true timeframe of canine companionship so difficult. We, as humans, have worked with canines for so long that it predates most written language! With agricultural animals such as sheep or hogs there are at least remains found from ancestral farmlands to date back their timeline. With wolves we have to do the best we can with the only remains that have been found thus far, and then base our findings around the era of hunter-gatherers of this time.

As a new companionship formed between man and canine, so too did the need to breed the canine species for different traits. Some people were looking to their new canine friends as a means to hunt with. Others wanted them as protection for their clans or villages. Whatever the reason, the first dogs started to come about during the Ice Age. Initially, it can be argued that the first types of dogs bred were more of a wolfdog mix.

While the first dogs were possibly from a relative of the wolf, as mentioned previously, they still would have shared nearly identical traits to that of the modern gray wolf. With this being said, the era of breeding out wolfdogs could have been more complex than a mere transition from wolf to dog. Wolves are wild animals, plain and simple. Just because you can get a wolf to follow your commands does not mean it's been fully domesticated.

Through selective breeding which most likely occurred during this time, humans were trying to get specific traits from individual wolves they revered. This would ultimately lead to changes in their DNA and the overall composition of the canine itself; much like how we selectively breed dogs today and wind up with a multitude of breeds. Over time, the wolves would become more dog-like in appearance and even behavior.

However, the transition from a wolf to a dog is crucial. The most common complaint about wolfdogs is that they are full of energy and temperament issues. They are stuck in a phase where they are not certain if they should be wild or domesticated. They can be known to run amok during this time of transition. Their brains are constantly battling their instinct to be wild while becoming domesticated, and they often lash out in this regard. One minute they could be perfectly calm and collected and the next they can tear up, destroy objects or act aggressively towards their owners.

It is uncertain if the wolfdogs of this time period acted this way. Naturally, they didn't mix a wolf with a dog for breeding; there were no actual dog species that existed quite yet. There is some evidence to suggest that these wolfdogs existed before human interference ever began. Some of these wolf remains had less teeth and shorter snouts, more similar to that of a dog. These wolves were undergoing physiological and psychological changes. But with any change in their structural DNA I am sure that the humans of this era faced several challenges in getting these canines under control.

Their adaptation can be delineated from multiple cases of genetic evidence. It is heavily argued that the domestication of the dog happened several times throughout the course of human history, rather than a single occurrence in their adaptation phase. It's hypothesized that dogs initially emerged in Europe, and again in Asia independently from the West. Also, there were two major events in history where dogs originated. One was from the domestication of wolves, and the other was from breeding a wolf-like relative en masse.

Wolf populations began to split apart between Europe and Asia and formed two distinct groups during this era. As they were splitting across continents, a common ancestor to the wolf may have been what evolved into the modernized dog. This is hypothesized because ancient dogs did not have characteristics that identified with the two wolf groups, which separated between continents during this time. The largest debate against this, though, is that there has not been any fossil evidence to prove that there was such a species; merely speculation.

Ancient dog remains do not have a resemblance to modernized dogs. There have been findings where ancient dogs had a mixture of wolf DNA in them, thus supporting the claim that dogs descended from wolves. On the other hand, dental paleontology suggests that the teeth and skeletal structure of ancient European dogs are much different than that of a wolf. These dogs most likely formed from another species of canine all together; as to follow an evolutionary line. With modern DNA testing, a sample of nearly 150 prehistoric wolves, wolfdogs and dogs have all been traced back to a single lineage of ancient gray wolves.

These genetic tests were proven to be highly accurate and were traced back by four lineages of the aforementioned wolves. One of the most supported theories currently out is that dogs did in fact descend from a specific type of wolf. But that wolf species has gone extinct and the current gray wolf is not directly related to dogs of today.

These arguments continue to go back and forth for an extraneous amount of time, and much more than I would like to get into for simplicity's sake. So to make things easier for everybody, we will leave this open for interpretation. Perhaps both theories are correct, or perhaps we will never know the dogs' true ancestry. Realistically, this entire matter is of lesser significance as we already know that wolves and dogs are very much related to one another. Whether dogs are directly descended from wolves or are from a similar species of canine, they all share the same roots and history. In fact, a dog's DNA is only 1.8% different than a wolf's, making the exact lineage between the two arbitrary in this specific case.

What we do know for a fact is that hunter-gatherers were the first to domesticate which ever canine they originally came into contact with, and that is how we have the dogs of today. The full domestication process occurred approximately 14,700 years ago, with some debate stating that it was 36,000 years ago. The earlier prediction stems from undisputed findings, while the latter comes from speculation. So, dogs have been man's best friend for the majority of modernized human life, and we have shared a special bond with them over the course of time.

Dogs soon learned their role in the world as they began sharing a path with humans. It did not take long for dogs to coexist with humans and partner up with us, making them one of the first recorded domesticated animals. Wolves have a precise displacement with their temperament as they are pack-oriented and have the mindset to work together. This allowed the domestication process to happen relatively quickly during initial breeding and the ability to control the animal in general. Humans began selecting which traits they wanted, which included size and subordinate displacement. Dogs finally emerged after years of selective breeding. Soon, they became one of the most cooperative animals to ever exist and remain so to this day.

Some of the earliest wolves that were domesticated and bred into dogs were thought to have a genetic disorder known as *William's syndrome*. This disorder can cause learning disabilities, defined personality characteristics, defined facial features and several infamous health concerns. It also increases levels of friendliness and happiness towards others which is a sought after trait when humans began domesticating dogs. This is a great theory and explanation as to why dogs bonded with humans so quickly.

One extraordinary benefit that dogs obtained during domestication was the alternation from being an obligate carnivore to a scavenger-carnivore. This is essentially the same as an omnivore, but with having a preference (and a requirement) for meat over other food selections. This could have been a forced dietary change with hunter-gatherers having to eat whatever was available to them at the time. Even so, many fruits and vegetables have proven to be extremely beneficial for dogs and help with an array of health related issues. This can also be seen in select breeds of dogs where some of them will eagerly beg for fruits and vegetables while other breeds want nothing aside from meat.

The dogs' adaptation phase also allowed them to digest starches and agriculture-based products. Just like the consumption of produce, the genes that allowed a dog to intake these particular foods were through an adaptation process based upon necessity. If meat was scarce yet other food sources were readily available, the dog would in turn adapt with whatever food was present.

We can even look at modern dog breeds and compare their genetics to ancestral canines that originated in the same locations of the world. Their dietary history matches up with both ancient and modern species digesting the same types of foods. Wolves have very few genes to digest vegetables and none to digest starches. This compares to dogs that are found in more frigid areas of the world which have similar genetics as wolves. They are not able to digest many of these agriculturally-based foods. In contrast, dogs that have been bred in warmer climates where agriculture has expanded now have a multitude of the genes to digest fruits, vegetables and starches.

Specific regions of the country where dogs have been domesticated and bred are more likely to digest these types of foods. There is an ongoing study which is based on a very precise species of wolf called the *Taimyr wolf*. To summarize, these were a group of gray wolves that shared a very similar lineage and an association with several modern dog breeds. They were relatively docile and preferred the company of dogs. These wolves were more than likely bred with ancient dogs which created certain dog breeds we see today.

The physiological appearance of the Taimyr wolf can be seen in modern dogs that were bred by the Inuit's and other settlers in higher latitudes. This is why you will see the Siberian husky, Alaskan malamute, Greenland dog, Eskimo dog, Samoyed and most other sledding dogs with a more wolf-like appearance. The introduction of this wolf-to-dog descent dates back over 15,000 years. It has helped scientists confirm that dogs have been domesticated in the new world longer than initially predicted.

European settlers and American natives began breeding dogs for temperament, and their passive traits were carried down dependently through this process. Humans of this era would hand pick obedient and loyal dogs to serve as a gateway into domestication. Once the temperament got to a deemed level of acceptance by humans, they started breeding these dogs for other traits. Some of the desired traits were for coat colorization, or an overall mutation in color that they found astonishing. Other traits were for tail and ear size, body mass, obedience and so forth. A few traits were genetic mutations that were unintentionally caused by selective breeding such as brachycephaly (short snouts). Some dogs also began having floppy ears instead of the ears being pointed, which was a revered trait for its time and is still carries over today.

This type of selective breeding process was recreated in the 20th century across the globe to create more of the breeds that we are familiar with today. It also helped prove that these historic renditions of breeding were not only successful, but that they could be done with any domestic animal henceforth. Thus, selective breeding continued to grow well throughout the 20th and 21st century.

The more that dogs were bred and domesticated, the further away they resembled their wolf relatives. As time went on and dogs became more attached to humans, an increase in their cognitive capabilities was also noted. They underwent a transformation that had never been seen before in any other animal. Their synapses started to alter, and dogs became one of the first animals to have an increased disposition in memory.

Dogs were able to learn verbal and nonverbal commands relatively quickly, and were able to help humans with daily tasks that were not thought possible for animals to complete. Dogs converged with humans in such a way that they can now pick up on our emotions, tone of voice and even facial expressions. They know who their owners are and form a bond with humans that surpass any other animal. The nonverbal cue of pointing has been used for several millennia to bring a dogs' attention to specific objects or areas of interest. There are entire studies based on the pointing technique that dog trainers use.

As this new revelation was bestowed upon early breeders, many of them found that dogs could carry out specific jobs to help humans in various regions of the world. There were some breeds that were so important to human development that their specific roles are still utilized today. To start off, we can look at some of the initial jobs dogs had that are still in use.

There were four critical jobs that dogs have served throughout time. We mentioned earlier that sledding dogs developed through the specialized breeding of the Taimyr gray wolf. These dogs helped people get across the tundra and mountainous ranges with a lot more ease than humans could have hoped to accomplish alone. There were also the notorious hunting dogs that were primitive for their time and extremely vital to the longevity of early humans. For agriculture assistance, there were herding dogs which kept animals such as cows and sheep in confined places and helped protect them; as to avoid them wandering off from the herd or being attacked by other animals. Lastly, there were guard dogs which were bred later on in history, and these became very popular during the Roman Empire.

Dogs that were bred for these specific roles constantly aim to please their owners in regards to their said role. They need to fulfill their tasks, and their brains are structured around doing so from years of conditioned behavior for that specific purpose. This is why you will see a lot of these working breeds with high energy today. Their jobs require ample stamina and extensive labor. There are people today who want these breeds and merely obtain them out of an admiration of how the dog looks. They do not take their behavior and energy levels into consideration. Therefore, the dog can become destructive or ill mannered in the household and will not settle down unless forcibly trained to lose that derived role. I am not blaming any owner for wanting to own a specific breed; simply stating that it's something to watch out for when selecting a working breed.

Now leading into the modern era, dogs are bred for a multitude of reasons. We still keep the job-oriented breeds that our ancestors used for their working dispositions. But there are other breeds of dogs that have been created through a selective breeding process for various roles. We now have a lot of agility breeds used for sporting. There are show dogs which are used to display their coats and precise characteristics that must meet specific standards to win tournaments. Then there are dogs bred merely for companionship, with no true role other than to be our friends. This is still a large role in our human lives today. And eventually we bred dogs smaller and smaller until we got to a toy breed size, which also fits into a companion-based breed.

Humans are still breeding dogs to get specific behaviors, sizes, coat colorations and a plethora of other traits. There are more breeds of dogs today than we ever thought possible. In fact, dog encyclopedias are being revised annually to add in newly acclaimed breeds that have met worldwide standards.

Dogs have hit a pinnacle in their involvement with humans. But what has happened with wolves? Both species are very closely related to one another, yet they are now portrayed in a vastly different spectrum in the modernized world. The next sections will go into detail about the obstacles wolves have run into and how they have adapted to humans over time.

3

Developmental Stages

Wolves develop fairly differently than dogs when they are puppies. At two weeks of age, wolves will begin opening their eyes and will start moving around on their own. They start to learn about the world around them and the basics of survival before they are a month old. Around three to four weeks of age, the pups will feel comfortable and confident enough to leave its safe haven, typically a den, to go explore the vast and open world around them. After approximately one month they are weaned off their mothers and will start eating meat with the rest of their pack. Sometimes the mother will regurgitate meat and feed it to their young until they are strong enough to move freely on their own.

As wolf puppies develop, they begin to understand the importance of a pack and what all it entails. The pup will figure out the roles that every wolf plays and how each pack member looks out for one another. At the age of four to five weeks, the pups will lose their baby teeth and develop their adult teeth. At this time, they are now able to go hunt and become involved with the pack. Adult wolves will almost always allow the pups to eat first, giving them the nutrients they need to grow over the short period of time in their developmental stages.

Pups will learn about their fear stimulus in correspondence to certain factors within the first month of life. This is what triggers the fight or flight response, and is also the reason why wolves do not approach humans very often in the wild. They have an uncertainty or fear towards us. The fear stimulus differs between North American gray wolves and eastern gray wolves. The eastern gray wolf found predominately throughout Europe and Asia does not have the same anxiety stressors as the wolves in North America. This could be due to their exposure, or lack thereof, with humans in their natural environment. But eastern wolves have had a longer history of interacting with humans and occasionally being aggressive when confronted by them. This contrasts the behavior of North American wolves, where they are generally more timid and reclusive around humans, unless they have been exposed to people very early on.

This brings us to wolves raised in captivity or living on sanctuaries and reservations. It's important to understand that wolves today cannot be domesticated. They are merely able to be tamed. There is a big difference between these two terms. Domestication means that the animal has a symbiotic relationship with people, a relatively docile temperament and can be kept as a pet or as part of livestock. An animal that is tamed means their aggression, anxiety or fear response is gone and will tolerate being around a person, but they are still considered to be wild animals.

Developmental biologists have worked with a few wolf pups over the years to get them to behave more like a dog around people. Their fear synapse dissipates towards humans whenever they are reared by people, and they start to act much friendlier or open towards everyone. Wolf pups love playing similar to how dogs do, and can acquire the same enrichment benefits as their relatives. Some wolves are even given puzzles for enrichment that they must solve cooperatively in order to receive food or treats. While this may be a breaking point in the theory of how wolves were initially domesticated thousands of years ago, they require constant care and attention to behave in this manner.

Wolf pups playing with a camera strap,
Wolf Creek Habitat

Wolves on sanctuaries also become more docile around humans and even depend on them for survival. They know where their food sources are coming from: humans. A number of wolves on sanctuaries around North America will allow their caretakers and visitors to come up close and even pet and play with them. While this is a remarkable event to experience if ever given the opportunity, it's important to realize that the wolf must first be raised and exposed properly to behave in this manner. A person cannot simply buy or capture a wolf and expect it to act the same way it does on a sanctuary.

Dogs do not share the exact same developmental stages as wolves do. They will open their eyes around three to four weeks of age, which is a week or two later than wolf pups. The majority of a dogs' early life is learning socialization from both their mothers and their human owners. Puppies are weaned around eight weeks of age and are placed on puppy foods and formulas vital to their growth. It's important to feed a puppy specific food to their age, as these foods contain a chemical known as DHA. This chemical replicates what is found in a mother's milk for brain development. They should be kept on these types of formulas until they are roughly a year old.

Albeit a little disgusting, there is an accredited theory as to why puppies eat their own excrement. It's likely due to them not receiving enough nutrients from their diet, so they try to reabsorb any that did not get properly digested. Wolf pups do it for this exact reason, so it's a safe bet that dogs inherited this behavior as a way to keep their nutrition levels intact.

Enrichment is a crucial component to a dog's development. They need something to entertain them as they get bored just like we do. Something for them to chew on is a great source of enrichment at an early age while they are cutting teeth. Their teeth will grow in around the same time as a wolf's, which is at the four to five month mark. Various toys are used to keep them busy and to work off some of their energy. Depending on the dog, they can be hit or miss with certain toys such as balls and Frisbees. It all depends on a dog's personality and what toys or material they like playing with or are exposed to early on. Playing fetch is a great way to have working dogs use up a lot of their stored energy throughout the day.

If dogs know that treats are involved, they are more likely to behave in ways that you desire through positive reinforcement. This is how you can train dogs to do tricks or even used as a way to reinforce good behavior… such as going to the bathroom outside rather than on your carpet. Nearly all dog trainers will use small treats to reward dogs whenever they have done something that is desired: obedience behaviors, tricks, listening to commands, etc. Eventually the dog will learn how to do these tasks without treats based on perpetual reinforcement. This is what is known as Pavlov training, a term coined by behaviorists and psychologists. It works on other animals as well such as primates, felines, cetaceans, and even fish to name a few.

Unlike wolves, dogs do not learn the same fear

responses at an early age due to how they are domesticated. Dogs' synapse reception is a little different and they grow accustom to humans helping them out whenever they are in trouble. If you have ever owned a puppy, you probably saw your dog looking back at you for help whenever they couldn't solve a physical problem to get to something they wanted. They continue this practice throughout adulthood due to the relationship they have with us. If a dog was never raised around humans, though, they become more goal oriented and will persist on getting something that they want instead of seeking help. This behavior is seen in a lot in animals that are in pounds and even coincides with how wolves behave.

Throughout the thousands of years of domestication, dogs have learned facial expressions of humans rather well. They are able to tell whenever we are upset, sad or happy about something and will even go as far as to mimic us. A new study has found that dogs have the ability to copy a multitude of facial expressions that humans make. Dogs do this in order to get something they want, play with us, get out of trouble, go outside, and so on. They know how to play us and will give us those puppy dog eyes consistently to get what they want. Whenever they know that they have our attention, they will raise their eyebrows, stick out their tongues or vocalize with us. Try observing this for yourself the next time a dog comes up to you.

My Pembroke Welsh Corgi responding to the pointing technique

A person pointing to an area of interest is one of the most complex things a dog has ever learned. It is constantly being studied and expanded upon in its use. A dog will look at a person and then to the general vicinity that their owner or whomever is pointing towards. It takes them a few tries before they know what it is that the person wants in that direction, but they learn how to pick out objects from the area that's being pointed towards. Eventually they will understand what humans are asking of them and will even learn words that are used in conjunction to pointing. This is a way that hunters train hunting dogs to find game. These types of dogs must be trained at a very early age in order for it to work, but it is outstanding to see in practice. The knowledge acquired by dogs from the single cue of pointing is something that not really any other animals have learned to comprehend.

The expressions and cues practiced by humans are a significant way that dogs separate themselves from wolves. The wolf has never really learned what to look for in a human as they have always been wild. Even in captivity, wolves still don't fully understand our cues. If accompanied by dogs, they only pick up on what the dogs are doing for attention or for food instead of what the human wants them to do.

If you are raising a puppy from an early age then it's important to understand that dogs can obtain psychological issues. I had to experience this first hand when my dog developed separation anxiety since I worked full time when he was a puppy. Not all dogs will get anxiety in this way, but many of the ones I've personally seen will get anxiety during their developmental stages after being left alone with no parental figure at home to support them. Signs of anxiety can range from the dog chewing on its paws, obsessive licking, illogical barking towards you or tangible objects, sporadic temperament, and lethargy. Anxiety is the biggest problem that dogs face in a neurological sense. Though, they can also develop psychological issues such as depression, compulsive disorders, and excessive fear towards a particular noise or people.

Wolves also develop these psychological disorders, though not as commonly as dogs. The most common disorders for wolves are fear, which keeps them alive and away from stressors in their environment, and depression which is generally caused after a member of their pack has passed away. The howl of a wolf that has experienced the death of a fellow pack member is one of the saddest, most somber noises to hear. Wolves that lose their alpha leaders become lost emotionally and are unsure of what to do for a while.

As more wolves are hunted throughout the United States, these neurological problems start to surface and can be seen both in wild wolves and those relocated to sanctuaries. It can cause the wolf to act erratic, scared or even intolerant of humans. They have a harder time coping with the loss of pack members, much like we do with friends or family. To them, the death of a pack member is equivalent to losing a family member, if it wasn't a family member that died in the pack to begin with.

Wolves require their parents to be around at all times during the early stages of life. They need to understand their role in the world and how to adapt to their environment, as harsh as it may seem at such a young age. But wolves are very social and intelligent animals. Every member of the pack supports each other and stays together to hunt and travel. As wolf pups age, they begin to understand their purpose in this world. If not for the wolf, prey populations would run rampant and destroy much of the greenery and water reserves in nature. Wolves help to balance this and keep herbivorous populations in check.

While dogs do not share this same role, they do serve the purpose of fulfilling the needs of humans as their companions. After years of domestication, dogs have learned to rely on us even at a young age. Therefore, the most important step in raising a puppy is to show affection and commitment towards them. Dogs look up to us as family figures and rely on us to help them with everyday tasks that they cannot accomplish alone. Raising a dog is very similar to raising an actual child. They test you, wake you up in the middle of the night to go to the bathroom, want to play with you and can even seem needy at times. But the number one thing to realize in their early development is that they need you as a part of their lives. They will show you loyalty, providing you nurture and care for them.

4

Facing Endangerment

Wolves have suffered and endured many hardships over the course of history. For millennia they have adapted, evolved and faced what many animals could not have possibly withstood. Their survival has been built upon working together not only with their own packs but also with nature itself. Yet wolves are now facing a dilemma that they perhaps were not prepared for. This quandary is with the adaptation to human civilization.

As humans began to expand their towns and communities northward, wolf sightings became more frequent. The gray wolf is naturally found in mountainous regions and areas with higher latitude. Therefore, it was only a matter of time before American settlers started invading their homes. The regions of the gray wolf's original domain in the United States and Canada are considered the Great Plains and the Woodlands. These regions for much of the 18th through 20th century were considered wild; some parts still are to this day. Native Americans were the only ones to dwell in these areas for a long time. As the land was considered wild and free game to everybody, hunting was the norm. Many predator species such as wolves and bears were targeted because they could potentially limit the settler's food supplies through competition.

The beginning of the 20th century was a dire time for gray wolves. By the 1920s many of the wolves native to these lands had been wiped out. Even though Yellowstone National Park was founded in 1872, it did not offer any sort of protection for the gray wolf. Not only were they vulnerable with no mandated ordinance to protect them, the United States government legalized the culling of all wolves from the northern states for the general sense of prosperity among hunters and farmers. They were hunted relentlessly and people were even given bounties for every wolf they killed. The last of the gray wolves within Yellowstone were eradicated in 1926.

As time went on, detrimental effects from the lack of predator species began to surface in the park. Rivers were drying up. Areas that once had plentiful vegetation became blighted. The aspen and pine trees started to wither. The causation of these problems all stemmed from an over abundance of deer and elk. They were razing Yellowstone since they had very few natural predators left to hunt them. Hunters weren't even able to balance the myriad of deer and elk out as hundreds, if not thousands, were born annually. This went on well into the 1980s as the land worsened.

Something had to be done about this, as the park was in ruin. The state and federal government banded together to help resolve this dilemma. After careful review and consideration, both parties agreed that wolves must be brought back to restore order to the national park. January 12th of 1995 marked a glimmer of hope back into Yellowstone. Eight Mackenzie Valley gray wolves from Canada were captured and relocated to Wyoming. They were set free in the greater regions of Yellowstone in hopes to restore balance to the park. And balance is exactly what they brought.

The pack of eight started reshaping parts of the national park, but much more had to be done in order to revitalize what had been lost. Biologists proposed that there must be a significantly higher number of wolves in the park to bring down the deer and elk populations. The following year, twenty-five more wolves were set free in Yellowstone. Wolf populations began to increase as pups were being born annually, and the elk and deer populations went down to an acceptable level. With this transition, rivers started flowing back to their original levels. Trees and vegetation rekindled all over the park. Yellowstone started to regain some of its natural beauty, all thanks to wolf involvement.

However, with so many wolves being born every year there were some challenges that came with it. Wolves do not know human boundaries that are administered to them; the wolves were only protected within the park and not outside of it. Naturally, wild animals are going to wander outside of these invisible jurisdictions. Once they did, some of the wolves started attacking farmers' livestock. This caused uproar in Montana, Wyoming and even Idaho. Hunters and farmers took to arms and started killing off the wolves that got too close to farms and towns.

This has been an ongoing problem between wolves and humans to this day. Many of the wolves are now being tagged or given tracking collars for biologists to keep tabs on. This is to help cease their encroachment and offer protection before they are killed. This doesn't always happen, though, and some hunters kill the ones wearing collars despite it being illegal to do so. The parks department and the states involved have been going over protection doctrines for several years to find a viable solution to wolves outside of protected jurisdictions.

As unfortunate as it is to see a wolf die, farmers with legitimate concern for their livestock have no other choice but to shoot or trap them. What I mean by legitimate concern is that some of these farmers are telling the government that wolves are killing their cattle or chickens even though it's not necessarily true. The state will compensate the farmers if they can 'prove' that a wolf killed their farm animals. A few of these farmers, at least in the past, were exploiting this and setting up false accusations to receive accommodations from the state government. This in turn also led several lawmakers to believe that wolves are more problematic than they actually are.

People began taking this wolf scare into their own hands and setting up traps around their houses and farms. Most of these traps were snares, such as bear traps, in order to keep wolves away, and to kill the ones that encroached on their farmland. A couple wolves did fall victim to the traps but not many. Once wolves knew what was going on they steered clear of many residential areas if they could help it.

It turns out that the wolves were not the ones getting snared in these traps. Rather, dogs owned by local residents were becoming ensnared in them. Nearly 70% of animals that were falling victim to these bear traps were the dogs around town. This instantly led to public outcry and many farmers began disassembling the traps. While I do sympathize with farmer's losing some of their livestock, I was also relieved to learn that wolves were no longer becoming entrapped in these barbaric snares and left to die.

I would like to point out that the gray wolves of North America are not endangered, aside from the Mexican gray wolf. Only specific regions such as Yellowstone, Montana, Minnesota, Oregon and California have wolf populations that range from threatened, vulnerable or endangered. This is not to cause confusion thinking that all gray wolves are enlisted as a protected species under the Endangered Species Act. While many wolves are not found in the contiguous United States anymore, the majority of gray wolves today can be found throughout Canada and northern Alaska. Roughly 95% of gray wolves in the United States have been exterminated over the course of human expansion.

I took a trip to Jackson Hole, Wyoming and Yellowstone during the winter of 2016. My goal for this trip was to photograph gray wolves in their natural environment and gather as much information as I could about their involvement in northern Wyoming. For the life of me, I couldn't find wolf tracks or sightings of them anywhere. This wasn't just me aimlessly driving around and hoping to find them either. I do this professionally and would like to think that I know how to track animals to a degree. I mapped out exactly which areas they would be in at this particular time of year and went early in the morning to scope them out. After hours of taking off-road trails, roads covered in ice and routes leading up mountain passes, I could not find a single wolf. Not even the footprint of one!

One of the national parks guides whom I traveled with the following day told me about hunters trying to kill wolves off or scare them away with firearms. He mentioned that this apparently worked at first, but once again wolves caught onto human interaction relatively quickly. The guide snickered at question as to why I couldn't find any wolves despite this occurrence. His response was, "Oh yeah, you're looking in the right places. But the wolves around here caught onto humans neutralizing them so they changed their sleep schedule to avoid us. Visitors are lucky to ever see one around the late winter season anymore… especially during the day."

While I was disappointed about not being able to

see any wolves during my stay, I was taken aback to learn this important piece of information. Wolves that came down to the Teton Mountain range learned within months' time what humans were capable of. With this information, they went back up to Yellowstone and completely switched from being diurnal to nocturnal to avoid encountering humans. This opened my eyes as to how intelligent wolves actually are and what they're capable of as a pack to ensure their own safety. The whole pack didn't even have to learn about the trapping or hunting that was going on. All it took was several wolves fleeing the Teton region from gunfire and reporting back to their packs. This is not necessarily to say that they're all safe from hunters, but merely that they were able to communicate with each other about what times of the day their pack should go out and hunt.

Northwestern wolves such as the Mackenzie Valley and Timber wolves of Canada, and the Tundra and Great Plains wolves of Alaska have had their share of problems as well. While hunting is still legal with wolves in these areas, there is infinitely more land for the wolves to withdraw to. This means that if wolves catch on to humans killing off some of their members, they can at least retreat to higher elevated areas or in the remote wilderness. Much of northern Canada and Alaska is unscathed by human interference so it acts almost as a safe haven for them.

The problem branches from wolves in the southern Great Plains and Woodland territories. They have been known to unintentionally encroach on human villages and two reported attacks happened in the last 60 years. The ones that did attack in these instances and prior to this time frame were more than likely infected with rabies, which can cause irrational aggression. This caused hysterical outrage among the local residents for decades. Many of the wolves were culled from these southern areas, and sightings of gray wolves diminished significantly; although there are still a few packs to be spotted.

Timber and Mackenzie Valley wolves are not similar to the wolves we would typically see in the contiguous United States. They are the largest species of gray wolves and can grow up to 135 pounds (61 kilograms). Coming across a rabid or aggressive pack would frighten anybody. Wolves of this size and caliber have enough force behind their jowls to apply over one thousand pounds (454 kilograms) of pressure per square inch to their target. This type of attack by one or two wolves alone is enough to take down a 1,400 pound (635 kilogram) moose. It's no wonder that attacks unto humans caused such a massive panic.

Mackenzie Valley Wolf,
Colorado Wolf and Wildlife Center

Hunting wolves in Alaska almost became the norm for several people in the state. The first goal was to remove them from human territories which more or less had been accomplished. Wolves became extremely fearful of people over time in particular regions. Though, hunting didn't just stop there. Acts were put into place that allowed the killing of wolves in order to set up oil reserves around the state. Some of these reserves weren't anywhere near wolf encampments which caused mixed emotions from state citizens. Other claims were to protect moose, caribou and deer populations from predator species. Bounties were then put on wolves' heads by the state to eradicate them out of specific areas. Helicopters started taking hunters over these regions where they could begin sniping wolves.

I'm not going to divulge into the moral and ethical standings of this act. But I will say that this ruling was overturned in 2016 by the presidential cabinet. It banned helicopters from being used to kill any predator animal in Alaska, under almost all circumstances. It also restricted the generalized hunting of wolves and bears in wildlife refuges. Wolves became safe from imposing threat, and education groups were set up to teach the public about wolf conservation in these areas.

Then, all of a sudden, this bill was abolished in 2017 under the new presidential administration. The Endangered Species Act is also at risk of being dismantled due to lobbying and a presumably potential confliction with states' rights. It seems that there will never be a resolve to wildlife rights in the United States. Laws are constantly changing or being ratified to either protect or endanger animals around the country. This is not to say that other countries are not doing the same thing. Wolves in Canada are still facing an impasse with hunting and conservation rights. But living in the United States, I am able to see firsthand how we may never have an established doctrine for truly protecting an animal, despite their conservation status.

The Mexican gray wolf has faced oppression worse than their northern relatives. These wolves share many of the same characteristics as the northern gray wolves, except that they are slightly smaller in size and have a lot more yellow and gray to their pelts. Most of them don't even grow over 70 pounds (32 kilograms). Their native homes used to be in Mexico, Texas, New Mexico and Arizona before their population began to dwindle. Many areas still refer to them as *lobos*, the Spanish name for wolf.

Similarly to the gray wolves in Yellowstone, people began hunting the Mexican wolf to completely eliminate it from the wild. The causation was primarily geared towards farmers protecting livestock, but many reports about the wolves killing farm animals throughout the years were either highly exaggerated or false. Wolves make up less than one percent of any attributed death to livestock, as the wolves themselves are fearful of humans.

The Mexican gray wolf now hinges between endangered and critically endangered. With that being said, they have become virtually extinct in the wild. Roughly 250 to 300 Mexican wolves are alive in captivity to this day. The only thing keeping these numbers stagnant is with the help of human involvement.

Concern of the Mexican wolf started in the 1970s when nearly all of the remaining species were facing extinction. The federal government and wildlife biologists teamed up to prevent their demise from happening. In 1982, the Mexican Wolf Recovery Program was established to help protect the species from going extinct.

The U.S. Fish and Wildlife Service now has the species legally protected, and conservation groups are focused on keeping the wolves captive to ensure their survivability. Many of these groups are working on rehabilitating the wolves in hopes of one day releasing them back into the wild. These groups, which are mostly nonprofit, along with local governments want to raise enough Mexican wolves to remove them off the endangered species list all together.

Reintroduction practices were implemented in the

1990s but they were met with failure. The problem faced with releasing them back into the wild was that a significant number of the wolves were either killed by humans despite it being a punishable offense, or they were recaptured by other groups. Currently there is a hold on releasing anymore back into the wild. Conservation groups want to make sure enough puppies are being born annually to drive their overall numbers back up.

Despite living in captivity, one upside to the wolves being held in reservations and sanctuaries is that their longevity increased from 3-5 years in the wild to 15 years under human care. Many people are happy that conservation groups and state governments are working together to safeguard this impressive species. Providing that no outlandish enactments are ratified, the Mexican gray wolf could one day walk freely in the wild without fear of being culled to the brink of extinction.

Red wolves are another threatened canid species that live in North America. They too have been hunted and exploited for decades, which has put them on the critically endangered list. Their original distributions were the southeast of the United States and had even been seen in

Louisiana and parts of Texas. Nowadays, they're an extremely rare sight around Florida, Tennessee and the Carolinas. They are so rare, in fact, that only an estimated 30 to 50 red wolves are still alive in the wild today.

The red wolf has a very close resemblance to that of a coyote. Their coats can range from an auburn red, black, gray, or a mix of all three which is much like that of a coyote's pelt. There is currently a reputable genome theory that suggests red wolves were a diverged species from coyote-wolf hybrids. It's believed that ancient wolf species began breeding with coyotes thousands of years ago which caused the manifestation of the red wolf to occur.

Red wolves are a very unique species because they are not related to the North American gray wolf. They are more closely related to European wolves in DNA and have a mixture of coyote genetics spliced in. This indicates that ancient hybridization between canidae members occurred much earlier than we initially predicted. Also unlike the gray wolf, red wolves are opportunistic scavengers. They will eat anything they can find in the wild such as vegetation, small animals and even insects.

After being relentlessly hunted for many years, the red wolf was a critical species to be saved by the United States Fish and Wildlife Department. Like the Mexican gray wolf, there are recovery programs established for the red wolf to ensure the survival of its species. There are roughly 200 red wolves kept in recovery programs around the country with the intent to bring their numbers back up. Reintroduction programs have been set up since the 1970s to allow some of the monitored wolves to be released back into North Carolina and Tennessee.

The fate of these canids is crucial to wolf conservation. They have quite a few differences from the standard gray wolf that are worth mentioning. Red wolves typically bond for life instead of seeking multiple mates. This is a rare trait among animals and is thought to only happen with some of the most intelligent of species. They have become a nocturnal species over the years, which have kept their acute population alive. They also use facial expressions, forms of howling and various body postures to exhibit their emotions toward a particular situation or other wolves. Their packs remain incredibly small if they choose to even form one. Such packs solely consist of the mating pair and their offspring, with no introduction of outside wolves into the pack. Hunting for food is done either with their tight knit packs or it is carried out solo.

Red wolves have become so secretive that it has been difficult for wildlife biologists to gather information on their natural disposition. These canines could open the door for many scientists to study complex social bonds for animals with higher cognitive abilities. They are so unique in the realm of canids that there is no other replacement species able to replicate their way of life. As of today, red wolves are considered critically endangered. The extinction of the red wolf would essentially mean the loss of hundreds of thousands of years of adaptation and evolution by a single species. The hope is to see this species thrive back in the wild where it belongs without the threat of being poached.

Any animal facing endangerment is never easy. Every species whether it is big or small, terrestrial or marine, aviate or burrow-dwelling can teach us something about life that we could've never known existed. To this day, we are learning more per species of animal and how they've adapted to their surroundings than scientists thought could have been possible. If we wish to observe how sustaining a healthy ecosystem naturally works, then wolves are the way to do this. Restoration of these canines is vital to keep prey species down. Without wolves, all animals that people consider nuisances or pests would decimate our crops, land and infrastructure faster than we'd be able to keep up with. Only nature can ultimately resolve itself in the long run. Removing a natural predator for the sake of prosperity and economic gain is not a way of life that anybody should consider.

5

Hybrids

As wolves began entering the lives of humans, they became of symbol of wilderness revered by people throughout the world. The more people began observing wolves in the wild or at sanctuaries, the more they wanted one of their own. This led to wolves being captured for trafficking. They were also used to breed with domesticated dogs in order to create wolf hybrids.

Once wolves and dogs intermingled in captivity, they started producing wolfdog hybrids which are unlike either of the two canines they stem from. Wolves are predictable. Dogs are predictable. Wolfdogs are not. Their temperament is erratic and can be borderline uncontrollable. More often than not, they start to become aggressive towards humans and even their owners. Wolfdogs share the apex predator role similar to wolves while remaining comfortable around humans like dogs are. This makes them dangerous. Wolfdogs resemble something in between both species and are confused about where they fit in. This isn't to say that all wolfdogs will act this way. I've worked with several that were docile and friendly towards humans. But there is a very good chance that many of them will show signs that have been mentioned.

Two wolfdogs interacting with each other,
Saint Francis Wolf Sanctuary

As it was brushed on earlier, the creation of this hybrid is essentially the same as restarting thousands of years of domestication and they can become impossible for most people to control. Those who acquire wolfdogs tend to relinquish them to wolf sanctuaries where they can roam free in several acres of isolated land, or they are put down. The breeding of wolfdogs is not only irresponsible due to creating an animal that almost nobody can control, but it's also out of ignorance since people think these dogs will become tame and have the generalized appearance of a wolf.

There is actually a mixed breed of dog that replicates what wolves look like (to a degree) and will retain their docile temperament. Breeders have been combining Siberian huskies or Alaskan malamutes with German shepherds to create a dog that looks more wolf-like. However, some people are exploiting this and try passing these mixed breeds off as real wolves. In doing so, the people exploiting dogs in this manner can get the animals euthanized since several states do not tolerate the ownership of wolves. If somebody ever tells you that they own a wolf, ask them if they have a DNA test to prove it.

A good way to tell a wolf apart from a mixed breed is that wolves will never have blue eyes unless they are newborn pups. Their eyes will always be amber or yellow in adulthood. Malamutes and Siberian huskies mixed with shepherds will more than likely have blue eyes from the lack of a dominant melanin gene. This is the easiest way to distinguish a dog from a wolf or wolfdog. One other distinguishable factor is that wolves do not wag their tails. They may move them from side to side in large, swaying motions but will never wag them like a dog. Their tails also do not curl; they stay level with their bodies or are in a downward position.

Coywolves are another hybrid animal that is quite uncommon to see. They generally only occur in the wild as you cannot really force two separate species to mate while captive. If they are in fact bred in captivity then reproduction must be done through artificial insemination. The interbreeding of coyotes and wolves is rare, though. Scientists speculate that this only happens if wolf populations are significantly low in one area and there is an abundance of coyotes around. The hunting of gray wolves is the primary reason these hybrids exist. Gray wolves in certain areas can be limited with mating partners due to eradication efforts. They resort to coyotes for mating and will share symbiotic relationships with them when forced to do so.

Their size is between that of a wolf and coyote, naturally. Packs of coywolves tend to be very cooperative with one another and allow other canines to play with them. Many of these hybrids are found in the northeastern United States and southeastern parts of Canada. Their coats are more auburn or brown compared to their wolf counterparts. They also howl much like a wolf but their vocalization changes halfway through to a coyote's yipping. It's very interesting hearing this in person.

Coywolves are closely monitored and studied in both the United States and Canada as they are a relatively new hybrid to emerge over the years. The inclusion of this hybrid into the world of canines gives hope to the return of wolf populations back up in certain regions. Some of the monitored coywolves are following coyotes throughout North America and are showing up in unexpected areas.

Red wolves are considered a hybrid species, even though they have been classified as a separate species of wolf. Their DNA dates back thousands of years and is a mix between that of eastern gray wolves and coyotes. They have been separated from other canid species for so long that they are now considered their own species of wolf. Even European eastern gray wolves are labeled as hybrids since they share a mix of genetics with North American gray wolves and coyotes.

Unfortunately, courts are trying to reclassify red wolves as hybrids rather than their own species. This is upsetting for one specific reason. The Endangered Species Act gives protection to purebred animals. It does not include any sort of protection for hybrids or mixed breeds of animals. Therefore, if they are labeled as hybrids in North America then they can be taken off the endangered species list and will no longer be considered critically endangered. Their protection rights would therefore disappear. Conservation groups have been arguing the inclusion of hybrid animals into the Endangered Species Act for years now. Hopefully something comes of it soon in order to offer protection for these canines.

Coydogs are the final hybrid species of canine to be registered in North America. These canines are typically a result of male coyotes encountering female dogs that have turned feral. Reports of these species have been around for nearly 14,000 years which were observed by Native Americans and the Inuit people. During this time period, coyotes and dogs were purposely bred together by the indigenous people to create hunting dogs and guard dogs.

Like the wolfdog, coydogs are just as ill tempered if

bred for domestication purposes. They have that inner turmoil within themselves and do not understand if they should be wild or tame. Many coydogs have been reported to act more aggressively towards humans than wolfdogs in captivity.

There is an uncommon breed of dog known as the Native American Indian dog. This breed is thought to have an ancestry and genetic makeup of both dog and coyote. These are very loyal and trustworthy dogs. However, they were bred by native peoples for hundreds of years to have this docile temperament.

Hybrids exist in many species of animals. As more land becomes restricted for wildlife to live on, the more common it is for species of similar genetics to begin mating out of necessity for the species' survival. Many of these hybrids are highly uncommon, so it's remarkable to see them appear in the wild. Their existence allows us to learn about how their temperaments and way of life differs from the two species that gave birth to them. Sometimes these offspring carry over many traits and behaviors from their parents, but other times they are completely different and in a class of their own. As more and more land becomes settled, destroyed and paved over, we are bound to see a greater number of hybrid animals pop up throughout our lifetime.

6

Wild Dogs

There are a few dogs in various locations around the world that have become truly wild. These species of dogs are quite interesting because they used to be domesticated at some point, but were either set free or escaped captivity. Afterwards, they started mating and reached a point where all of their characteristics became relatively the same within their species. Once their numbers had increased, these feral dogs started forming packs much like wolves do.

This became a curious situation in the realm of canines, seeing as these wild dogs' behavior was very similar to a wolf's. Wolves have a predisposition to establish complex social structures and group coordination to survive. They were the only canines to do this until wild dogs came into the picture. Once these dogs that had been set loose into the wild realized that humans were no longer a viable option for survival, they created their own packs and social structure. It helped provide explanation to an ongoing theory that an animal will revert back to its wild instincts if left to their own domain, rather than die off from hopelessness or lack of direction.

Through centuries of interbreeding, wild dogs became their own breeds rather than a conglomeration of breeds we see today. This means that they have their own traits, personalities, characteristics and dispositions that separate themselves from any other breed of dog. Rather than retaining domesticated traits, these dogs soon became a landrace species through their own adaptation phases and environmental circumstances.

The first wild dog that should be discussed is the dingo of Australia. Dingoes have been around for over 4,000 years and are thought to be related to the New Guinea Singing dog and the Basenji. They were believed to have originated from a wolf species known as the Tibetan wolf, which have much smaller legs than their European gray wolf counterparts.

During some point in the Pacific Island's history, the original dingoes were brought over by ancient seafarers onto the mainland of Australia. At this time, Australia was connected to Papua New Guinea by a land bridge. As singing dogs were from this country, it would further explain their relation to dingoes before they began diverging as separate breeds.

Dingoes are fairly different than any standard domesticated breed of today. Their muzzles, mouths and teeth are larger than many dogs, but their skulls are smaller in height. Similarly to the wolf, dingoes have a life expectancy that ranges between 3 to 5 years; this number drastically increases if in captivity. The biggest difference between dingoes and wolves, though, is that they are still aware of social cues from humans. Wolves never learned this obviously, but dingoes have an advantage of knowing what to look for in human behavior with gestures and facial expressions.

Dingoes establish packs just like wolves. The size of each pack is dependent on their location in Australia and the food and water reserves found in specific regions. Many dingo packs will range from 3 to 14 members, sometimes with multiple packs merging together to have a higher chance of completing objectives. Juveniles in most scenarios will leave the pack before their first year of age until they begin mating. Once they do, they will either create a pack with their mating partner or join an already established pack.

Dingoes try to avoid conflict with humans as much as possible. Many people with farms have hunted them to avoid their livestock being killed. Some animals that help deter dingoes away from farms are alpacas, donkeys and even other dogs. Poison bait was also used in the past to cull dingoes from human settlements, but this led to problems where domesticated dogs were getting a hold of the poison and dying. The use of poison has been met with conflicting reports amongst the Australian people for several decades.

Multiple instances have occurred where people tried to re-domesticate dingoes and keep them as pets over the years. While a number of reports have claimed they can be tamed for human companionship, they do not regain their domesticated disposition. Their mentality is still feral and they only abide by human authority because it's an easy relationship for them to fall back on. They get free meals and protection in exchange for being captivated. There is much dispute over the ethical approach to this, and it remains an open debate to this day.

The next wild dog that we'll be talking about is the painted dog of Africa, otherwise known as the African wild dog, terms I'll be using interchangeably. There are several distinct subspecies of this dog but they are all descendents from the same ancient breed. It's unclear how long the African wild dog has been around, but the first written account of this species has dated back over 1,800 years ago.

The DNA of the painted dog is fairly different than any other dog or wolf. Their teeth, especially the molars are larger than many dog species, indicating they probably descended from a common canid ancestor rather than European gray wolves. However, they do share more characteristics to the dogs of today than any other canid species on record. Many paleontologists believe that they share a lineage with jackals and Ethiopian wolves.

Painted dogs are very good hunters, work well in packs and share a lot of characteristics as wolves in this manner. They form packs not only to hunt but also to protect one another. The only difference between these wild dogs' packs and wolf packs is that the female is the dominant leader of the group for painted dogs. When hunting, the painted dogs will break off into two or three smaller groups while chasing their prey. These groups will begin flanking the animal that they're chasing and find an area to either corner or ambush it. They are able to successfully take down wildebeests and antelopes more often than not by using this flanking technique.

With their primary ranges being in Botswana, Namibia, Tanzania, Kenya, Ethiopia and parts of Chad, they must compete against other apex safari predators for food. This has resulted in a number of them being killed by lions that have invaded their feeding grounds. Hyenas also try to scavenge their kills while the dogs are still feasting, which results in tension and aggression between the two animals.

Packs of African wild dogs will do something interesting to let each other know when they want to eat. If the dominant member of the pack sneezes, then the majority of pack members must also sneeze to indicate that they should go hunting. This is thought of as a voting system, and if three to five sneezes accompany the dominant dogs' sneeze then they will go hunt. I'm not entirely certain how they realized to use this as a cue, but these dogs developed a pragmatic social structure to become aware of what synchronized sneezing meant.

A painted dog pack will range anywhere between 2 to 30 members. Unlike the dingo, African wild dogs do not tolerate captivity. They have been wild for so long with hardly any human interference. As such, humans in certain regions have found them to be a threat and have culled them off for centuries. This has led painted dogs to become an endangered species where roughly 5,000 of these dogs are still alive to this day. Luckily, they have a life expectancy much longer than many canines in the wild, living between 10 to 12 years of age.

The final wild dog species is the dhole of Asia. Dholes have been around for approximately 20,000 years which dates back to the end of the Ice Age. The first time modern humans encountered these dogs was in the late 1700s in eastern Russia. Their origin is unclear, but they are presumed to descend from an ancient species of jackal.

The dhole is actually thought to be the precursor to modern domesticated dogs. Their original range spread throughout Europe and North America thousands of years ago before they became isolated in Asia. Their isolation was most likely caused by the collapse of the Beringian land bridge. This supports the theory that humans at one time probably tamed this species before it escaped captivity or was relinquished back into the wild. They are the only species left in their genus.

Dholes are similar to wolves in several ways. They are incredibly social and form packs anywhere between 3 to

12 members. Sometimes they will form what is known as a clan which ranges between 12 and 50 dholes. This is typically temporary, where they will all band together to take down larger prey. They work as a team and will kill larger game such as deer and boars. Often times, dholes encroach on the territory of tigers and leopards which can lead to the death of pack members. Despite this, many dholes will work together to kill feline predators, and even bears with a high range of success.

The dhole is one of the few canine members that actively try to drive their prey into the water to kill it, as they are considered remarkable swimmers. If unable to kill larger game then they will settle for rabbits, reptiles or any small rodents. Dholes are able to run at 30 miles per hour (48 kilometers per hour) and flank their prey much like the African wild dog does.

Unlike wolves, though, they do not have any sort of alpha leader or hierarchy members. Every individual of the pack is responsible for the other members' safety and they communicate very well in this regard. Dholes also give feeding priority to pups before the older pack or clan members will eat. They are also not aggressive towards other dogs joining their pack, whereas wolves often do not tolerate members from other packs joining their numbers.

These dogs are short and slender in comparison to many dog breeds. Interestingly, dholes only have a set of 40 teeth whereas other dogs and wolves have a set of 42. They also have many vocal calls that are unlike any other canine species. Some of these calls are similar to the noises foxes make. While the use of their vast array of noises is unclear, it's speculated that dholes use these to coordinate and talk with their pack members. Some of these noises sound like chickens clucking, while another noise they make is similar to cawing. Although dholes are able to make a vast array of noises, they are unable to bark or howl like dogs and wolves, making them a very unique breed all together.

Dholes have been wild for so long that they are unable to be tamed by humans. There have been a few reported cases where dhole pups will interact and play with domesticated dogs, but will avoid other dogs once they are older. This is one species that has been relinquished back to nature and can never return to a life of domestication.

Unfortunately, dholes are now considered endangered with roughly 2,000 left in the wild. This brings their total population to less than that of wild tigers. They are disappearing at an alarming rate and are often overlooked by Asian governments as a species to protect. Hopefully with conservation efforts from various groups around the world we may one day see these wild dogs bounce back.

These three breeds of wild dogs are a very interesting aspect in the realm of canines. Each of them bears testimony of human history in regards to early breeding and where they were introduced throughout the world. We are able to learn a great deal not only with their ancestry but also how they behave in the wild. As they are all considered wild animals, their behaviors and personalities seem to resemble a mix of a dog and wolf. There is still so much to learn from these three species. Their livelihood could be vital to the research on roles dogs have played in early human development and domestication.

7

Symbiotic Relationships

S ome of the most social creatures on the planet build relationships and trust with one another. Psychologically, animals with a higher intellect and social disposition need to communicate with other fauna of the same family to feel satisfied. This helps them build assurance not only with themselves but also with other members of their group. Sometimes, though, communication isn't exchanged with members of their own family. It can be done with a separate species of animals all together.

The ultimate goal for a relationship is not to simply converse or interact with one another. Most of the time relationships between two individuals or groups are established out of a necessity for survival. This gets us into the discussion of symbiotic relationships. As mentioned earlier on, this type of relationship exists when two completely different species of animals will work together for a common goal. The objective between the two species could be for food, hunting, protection, guardianship, enrichment, or even to avoid an external conflict.

It may be of no surprise to you that wolves are animals that form symbiotic relationships in this manner. Several animals have had an accord with wolves for thousands of years. With that being said, some of these relationships have only recently been discovered. We, as wildlife scientists, are observing more about the wolf's way of life than we ever have in history. Therefore, some of these revelations about a wolf's social behavior have been astonishing to say the least. We'll now get into the relationships wolves share with various species and how they benefit from one another.

Coyotes

The relationship between wolves and coyotes has bordered the lines of love and hate. Both of these species are predators that compete for many of the same meals. As such, there have been conflicts between the two canines throughout history where neither are willing to back down if they're hungry enough. This has caused turmoil and even death between both groups when tension escalates. Some have even lurked around the dens of the opposing species which causes greater conflict. Canines are extremely protective of their young and will fight to the death for them.

However, not all interactions between the two species are always bad. In most instances, coyotes will follow wolves while they are hunting. The coyotes don't help in these attacks, but wait patiently until the wolves have finished eating. At this time the coyotes will come in and scavenge whatever has been left behind for an easy meal.

Wolves are aware of this and don't mind the presence of the coyote, providing they stay their ground. Some coyotes have been observed getting too close to wolves during feeding which will initiate the wolves chasing the coyote off. Neither party wants confrontation, though. The majority of coyotes will catch onto this and wait their turn for leftovers. I've personally seen coyotes follow bison around in Yellowstone as well. They keep to themselves as to not bother the other animals and walk beside larger species to forage or find carcass remains. The bison didn't mind their presence at all even though the coyotes were right in the middle of their herd. Following larger animals or apex predators such as wolves gives them protection as well as a meal for the day.

Coyote with a herd of bison,
Yellowstone National Park

There have even been events where wolves have bred with coyotes. This is what scientists believed spawned the red wolf of the southeastern United States. There are also areas in northern America where hybrid species have been spotted, having the DNA of both wolf and coyote. These hybrids are commonly referred to as *coywolves*, and have been around for nearly 100,000 years.

Dogs

The encounters between wolves and dogs are uncommon

and will generally only happen in a domesticated setting. Several wolf sanctuaries have wolves that do not mind the presence of dogs. They don't necessarily play or interact with each other, but they also don't mind each other's company. Some of the wolves I've seen in these types of sanctuaries will observe their dog counterparts and mimic their behaviors. They will paw at humans and whimper, copying what they see dogs doing to get food or attention.

The reverse is usually true in the wild, though. Wolves have been known to attack dogs if they feel the dog is encroaching on their territory. Smaller dogs that are vocal have the highest reported rating of being attacked by wolves. Larger dogs sometimes aren't even phased by wolves, and can stand their ground against them or don't try to cause trouble in the first place. Typically, the barking or yipping that smaller dogs emit is deemed a nuisance and wolves won't tolerate it.

There are rare encounters where wolves and dogs have formed friendships. One of these instances was with the wolf known as Romeo who lived in Juneau, Alaska. This wolf was very unlike other wild wolves. He would come up to people and play with their dogs throughout certain times of the year. Dog owners around this area would bring their pets around Romeo to socialize and play.

Even smaller dogs were safe around this wolf. Other wolves have been known to have similar interactions with dogs, so it's really dependent on the wolves' personality.

Dogs have also bred with wolves, but this is almost exclusive to situations that arise in captivity. This is why you will see some sanctuaries have wolfdogs, which they obtain through people trying to breed wolves and dogs together and can't handle their temperaments. Ultimately, they are then relinquished to sanctuaries or reservations to be taken care of.

Lately, people have been trying to purchase wolfdogs because they think it's cool to have a hybrid animal or something more wolf-like. I highly discourage the practice and the support of this. It took humans thousands of years to domesticate wolves into the dogs of today. Breeding dogs with wolves is the equivalent of restarting the domestication process. As discussed in earlier chapters, it causes a mix of behavioral issues in the hybrid species that are left unresolved. These issues can range from the wolfdog destroying everything in the house to becoming aggressive towards its owner or people in general. In the hybrids' mind, they don't fully understand what they are; if they're wild or domesticated. This causes them to lash out and act sporadically. Again, not every wolfdog will act like this but many will.

Ravens

If wolves had to pick a best friend then it would most certainly be the raven. The symbiotic relationship between these two species has been around for thousands of years. If northern gray wolves are in vicinity then ravens surely aren't too far away. This is the one relationship wolves have with another animal that is purely out of friendship rather than tolerance.

Ravens have had a history of helping wolves out in the wild, especially if the wolf is about to die from starvation. The raven will act as a wolf's eyes and scout out meals or carcasses of animals for the wolf to obtain. Ravens will make a distinct caw, signaling the wolf to come to a certain area, and the wolf somehow understands what the bird is trying to tell it. Other ways that ravens help wolves scavenge is by circling around a carcass until the wolf appears. Several documented reports of wolves on the brink of starvation were found to be aided by ravens searching for food for the wolf. These birds benefit from a wolf eating, too. They will scavenge the remains of which ever animal the wolf kills or helps break down the carcass of. Therefore, it is a win-win for both the corvid and canine.

But their relationship doesn't end with sharing meals. Ravens have been known to play with wolves and their pups for centuries. They will dive bomb and tug on the wolves' tails or fur to get the wolf to chase them. Once the raven sees the wolf stop, it will sneak up behind the wolf again and repeat this type of harassment. Wolves love chasing them around and never act aggressively in response to their playful tactics. It's fairly easy to tell whenever a wolf is angry or ready to attack such as raised hackles, growling, snarling their teeth and an erect tail. Not once are these behaviors ever expressed towards the raven.

There's some type of complex social bond between the two species which causes their desire to interact and help one another. Perhaps it's because ravens are arguably one of the smartest birds on earth, and wolves have a very high intellect as far as mammals go. Both species understand one another and wish to see the other party strive in correspondence to their symbiosis. Even when the wolves in Yellowstone were hunted to extinction and later reintroduced in 1995, ravens never forgot about them and exhibited the exact same relationship they once had upon the wolves' return.

Humans

Humans have had a mixed relationship with wolves over the years. First, we befriended them during the struggle of the Ice Age and tamed them to the best of our abilities. Then we began hunting and culling them out of certain areas in North America and Europe. Now many science and conservation groups are trying to save all wolf species from endangerment and extinction.

Wolves are naturally afraid of humans, probably for good reason. The more we've invaded their territories and disturbed their packs, the more fearful they became of us. If they sense that humans are around in the wild the wolf will try to avoid us at all costs. This is why actual wolf attacks towards people are so rare, despite the myths that have plagued wolves for centuries.

In spite of our rocky past with wolves, many humans still desire to help them. We have doubtlessly learned from our mistakes about poaching wolves out of natural habitats and strive to keep their numbers up in the present day. Luckily, wolves that are relocated to reservations and sanctuaries look for solace in us. They understand that many of us are there to help them and it shows by their complexion.

Whenever I work with wolf sanctuaries, many of the gray wolves act like dogs towards the staff and me. They will paw at us for treats and whimper if they want attention. The wolves were also perfectly fine with us going into their enclosures and playing with them. This is merely a trait that is restricted to captivity, though. But wolves in these scenarios are very appreciative of how much we do for them, which can be seen in their mannerisms and behavior.

Dogs are a completely different story when it comes to symbiotic relationships. They have been domesticated and tamed for such a long time that they are used to being around an excessive amount of animals. If I had to list out every animal that dogs get along with then it would be an entire book in of itself. So instead of doing that, I'll briefly talk about some of the unusual relationships I've seen between dogs and other species.

One of the most fascinating relationships I've seen is with dogs and cheetahs. My first time seeing this was at the San Diego Zoo. The zookeepers mentioned that cheetahs and dogs are raised together as infants because the dogs will help cheetahs socialize and become companions.

It also helps with the cheetah's temperament to behave in a more docile manner. The cheetahs at the zoo would constantly follow their dog companions around as if they were one of them.

Dogs have also exhibited playing and providing enrichment for seals in the ocean. The dogs I've seen will wade in the water towards the seals, and the seals will keep swimming laps around the dogs. They try to chase each other around in the water, the seal always winning of course. They both seem to enjoy each other's company a lot, despite neither species meeting prior to their engagement.

Foxes are another unusual friend of dogs. While they are both related to one another, foxes favor isolation from other animals. Some dog owners either adopt foxes as pets or have farmland where foxes come to visit. Most non-hunting breeds of dogs have been known to socialize very well with them. In most of these cases, you can see the dog and fox take naps together, eat together and even groom each other. Some of them will even chase each other around and pounce on one another in a playful manner. I should note that not many wild foxes want anything to do with dogs or any other animal for that matter. Their bonding is completely circumstantial and many foxes have to feel comfortable around humans and domesticated dogs to interact. With that said, foxes in captivity also tend to act very sporadic and unpredictable. It really depends on how early on somebody gets a fox and what they do to 'tame' it to prevent their wild behavior.

Relationships between animals can be very unpredictable at times. The most unusual of relationships are those that amaze us the most. For example, my dog loves playing with rabbits and gets really excited whenever they appear in the backyard. The rabbits in my area are used to him and don't appear to be threatened by his presence or his playful behavior at all. This is a rare occurrence because most dogs will kill rabbits regardless of the circumstance.

These types of interactions show that animals have their own personalities to them. Not every animal will behave in the same ways as other members of its family or breed. So it becomes very rewarding to see these types of relationships play out in the wild and even at home whenever you least suspect it.

8

Canine Diseases & Outbreaks

C anines have had a long history of health concerns. After modern domestication was set into place, a multitude of diseases began to manifest that had never been seen before. This was likely due to how various strains of diseases were prevalent in numerous animals, and we were simply not aware of it. The domestication process of wolves and dogs then introduced diseases into human populations. The outbreaks have not only affected dogs, but also wolves, coyotes, foxes and many other canids. Like humans, the bacteria and viruses present in canines evolve and adapt to new environments which makes it harder to combat. There is constant research and development with vaccines for dogs and even wild animals, and it is an ongoing process in the medical science field to ensure the longevity of these species.

If you own a dog, your veterinarian at one point probably went over the four lethal outbreaks that your animal should be vaccinated for. These would include rabies, distemper, heartworms and leptospirosis. I would like to point out that these diseases can occur in every canine species around the world. As a dog owner and somebody who works in wildlife science, I cannot stress enough how important it is to vaccinate dogs for all of these diseases. I have seen far too many animals suffer from these outbreaks, some of which do not recover at all despite our efforts to save them.

With that being said, there are several diseases that I would like to go over which are prevalent in nearly all canidae species. Each disease will be covered by their effects on the animal's body and immune system, along with their natural occurrence. I have worked with several DVM's to ensure that this list is accurate for dogs as well as wolves. I will not be covering every infectious disease. There are far too many to list out and it would be rather strenuous for you as the reader to skim through. To see a complete list, I would recommend looking at a veterinary handbook. The diseases will also be separated by those that solely affect canines, and those that are considered zoonotic which can spread from canines to humans. Some of these viruses do transition from animal to person and can be variably detrimental or even perilous to one's health.

Canine Diseases:

Canine Distemper Virus

This is a highly contagious disease that has no known cure. For dog owners, this virus is prevented by getting the DAPP/DHPP shot at a fairly early age followed by lifetime boosters that way the dog is protected over the course of its life. Canines in the wild do not have it so lucky, though. Inhalation will cause the virus to spread from one animal to the next, and can rapidly spread thereon. This virus primarily attacks the canine's brain and then moves onto the skin, pulmonary system, the gastrointestinal tract and the mucus membranes.

The animal will first run a fever and show signs of a flu or cold. Several weeks later they will have motor-neurological issues such as chronic shaking, uncontrolled drooling, and mouth quivering where it looks like they are shivering or chewing on something. After the virus transcends past this point the animal will start to wander around or seem confused about everything. In some cases the pads of their feet become rather hard or thickened to a near callous feeling.

Distemper has an increased prevalence in the wild where climates are tropical or hot. The virus is also able to spread with relative ease during heat waves and long summers. Climate change has been theorized to cause the increase of this virus for wolves and coyotes in the past couple of decades. Luckily, the virus is not very common with dogs anymore as many people put their pets on preventive care early on.

Adenovirus Types 1-2

Both of these viruses affect canines a little differently. Adenovirus type 1 is a precursor and the causation of hepatitis. It is highly contagious and spreads through contamination of bodily excrement or from being airborne. It begins to damage the liver, kidneys and circulatory system. Adenovirus type 2 is a respiratory infection that is spread in the same manner and is more common than type 1. Both versions of this virus can be prevented by a DAPP/DHPP vaccination for dogs, whereas wild canines have a slim chance for it to go away on its own.

The animal will run a severe fever initially if it contracts type 1. They will have a loss of appetite, suffer from colitis and can even vomit up bile or blood. Their eyes become very sensitive to light and the gums can swell and bleed. The skin of the animal can also turn yellow (jaundice) and have clouding over their corneas. Type 2 symptoms will be more similar to bronchitis and can cause the canine to have difficulty breathing.

Adenovirus will survive for 3-12 months on its own before it dies off in the wild. Most animals will try to move away from areas that are contaminated with this virus and may not return even after its eradicated. This virus is more common in North America, but there have been outbreaks of it in domesticated animals worldwide. Fortunately, type 1 has not been a major concern with wolf populations for quite some time. Though, the survival rates for animals that do contract adenovirus type 1 are significantly low.

Parvovirus

This virus should ring an alarming bell for most of you. Parvovirus is a dreadful disease which has a very high mortality rate behind it. This viral infection is so severe that it has a 50/50 chance of killing a canine that is undergoing treatment for it. Survival without treatment is slim to none. The virus works by splitting up cells in the blood, most of which happens in the gastrointestinal tract. This is why people who have seen parvo, such as me, can tell you that the smell is absolutely atrocious and very distinct. It has a strong smell of iron due to the fecal matter being filled with blood.

Aside from the intestines being infected by the virus, the cardiovascular system is also at risk. Canines exposed to parvovirus will become very lethargic, have vomiting and excessive diarrhea, and will refuse to eat. Signs to watch for are severe dehydration and the abdomen tucking inward due to pain. Puppies are more prone to getting this disease and it spreads rapidly from canine to canine. It's extremely contagious and can latch onto humans that come into contact with it (though, it does not affect humans in any way), which in turn will move onto other canine species. The vaccination of DAPP/DHPP is used to combat this virus in dogs before it ever becomes an outbreak.

Unlike many viruses, parvovirus cannot be eradicated by extreme temperatures. This is one disease that dissipates over time only, and will remain active between 5 to 12 months (typically an average of 7 months) in the wild or a home environment. This virus also sustains itself longer in wet, shaded areas. A single ounce of infected stool is enough to carry over to 35,000 canines that come into contact with it. Let that sink in for a moment to realize the severity of this outbreak.

Ehrlichiosis

This disease is also known as canine typhus. It is caused by tick bites which affect canines as well as humans. However, if a dog or wolf is inflicted with this disease then it does not transfer to humans or other animals; you must be bitten by a tick to get this infection. The most common way to treat this disease in dogs is with an antibiotic called doxycycline, which must be administered by a veterinarian. Ehrlichia is a worldwide outbreak and has been added to the CDC watch list from how common it has become within the past century.

Symptoms do not typically start right away, but rather a couple weeks after a canine is first bitten by a tick carrying ehrlichiosis. It introduces itself through the blood stream first and can then move to the liver, spleen and bone marrow. The lymph nodes of a canine along with the liver and spleen will become inflamed. This can cause the canine to become lethargic, lose interest in food, cause joint and muscle pain, shortness of breath and run a fever. If left untreated the disease will cause internal bleeding, impairment or inflammation of the eyes, and even neurological issues. Not all canines will have symptoms from this infection. Many have been known to carry it without any harm coming to the canine whatsoever.

A tick carrying this disease can remain dormant on a canine for a finite period until it is removed or voluntarily detaches itself. Some ticks will leave their hosts after several months, but this is more than enough time to cause the condition of the canine to become chronic or severe. Nearly every state in America has had a reported outbreak of this disease. So for dog owners it is best to keep your pet on a flea and tick prevention to help avoid this outbreak. I would like to note that symptoms for canines do not represent the same or similar symptoms for humans.

Heartworms

This parasite affects dogs more so than it does other canines. Heartworms are a type of nematode that lives in the atrium and right ventricle of a dog's heart. They are contracted by mosquito bites, especially during the summer months. The most prevalent areas that this occurs in are the southern states in America. It's much rarer to find at higher elevations and colder climates. In fact, those who live in the northern United States and in Canada have seen less than a 5% outbreak of heartworms annually.

A canine infected with heartworms may not show signs of having them at all. The mosquito transfers this parasite into the canine's blood, and the larvae will move its way into the heart tissue for several months. It takes around half a year for the worms to move into the atrium and ventricle of the heart. Once there, they begin to grow and become long and stringy in shape. At the adult stage of the worm's life, the canine can faint or go into cardiac arrest during exercise. The heartworms block the pulmonary arteries from receiving blood back into the organ which causes this to happen. The parasites can remain in the cardiovascular system for a few years after treatment occurs, and if left untreated the canine can die after long exposure.

Wolves have been prone to receiving this parasite but it's difficult to tell when they have it. Mosquitoes in the southern areas of America have been carrying this parasite for decades. Every dog that ventures outside in some form is at risk since mosquitoes are very common around these areas of the country. Heartworm prevention that you can receive from your veterinarian is the best method to prevent this from ever happening. A heartworm test is typically run before you can get the medication. If a dog has heartworms while being on the preventive medication, the nematodes will break apart in the dog's heart. When this happens, avoid having the dog do anything too strenuous.

The worms being eradicated out of the heart can clot inside of the pulmonary system and could cause the dog to faint or even result in death.

Lyme Disease

This is another illness caused by tick bites. Lyme disease is very common worldwide and is the most common tick-related illness in America. These ticks can be found anywhere, but their main areas are in rural northern states along with the Midwest. Luckily, humans cannot get this illness from canines; only if a tick latches onto the person.

If a canine is infected with Lyme disease then it may not present itself for several weeks, if at all. This disease is a bacterial strain that enters the canine and begins affecting the joints which then causes inflammation. The canine may experience pain when walking or have stiffness in their legs and dorsal area. Further symptoms include running a fever, lethargy, respiratory issues and a loss of appetite. If the disease becomes severe then it will start affecting the kidneys which can potentially lead to renal failure. In severe cases, Lyme disease can lead to heart and nervous system issues, although this has been reported to be rare.

The best method for preventing Lyme disease is

checking the canine for any ticks after going into wooded areas or places with tall grass. Animals in the wild are at risk of getting this bacterial strain, but there is roughly a 5 to 10% chance they will actually become ill from it. Treatment for dogs would include flea and tick prevention sold at many pet retail stores or by your veterinarian. Veterinarians can also remove the tick from its host if one happens to latch onto the dog.

Rocky Mountain Spotted Fever

Another very common tick related disease. This virus is more common in the Midwest of the United States than anywhere else. Like Lyme disease, Rocky Mountain spotted fever can only be transferred to humans if a tick attaches itself onto the host. Symptoms for this disease can take a long time to surface. Therefore, it's important that if you own a dog, you should check it for ticks regularly if you're located in regions where these ticks are prevalent.

The first symptoms to Rocky Mountain spotted fever in canines is running a severe fever that occurs roughly a week after exposure, along with lethargy. The canine will then have trouble walking, have swollen and inflamed joints or limbs, blood in their urine and will develop purple spots all over their bodies. If the virus becomes severe then it can cause heart arrhythmias, prevent blood from clotting, the canine going into shock or even resulting in death.

This disease is more common in dogs and coyotes than it is for wolves. The ticks often appear in spring through early fall, which is when they try to find a host to survive. Immediate medical attention is to be sought if your dog becomes exposed to this virus. Reported cases have been occurring often and many owners do not know their dogs have come into contact with these ticks until the symptoms become moderate to severe.

The following is a list of zoonotic diseases that canines can contract. Some of these infections can spread through bites, exposure to excrement, or by becoming airborne. An ample amount of these diseases are lethal to humans or can become lethal if left untreated. Medical help is to be sought if a canine you interact with has been tested positive for the listed diseases. It's better to be safe than sorry if possible exposure occurs under any circumstance. Many of these outbreaks can be treated without much harm coming to canines or humans, if caught early on.

Zoonotic Canine Diseases:

Leptospirosis

This is a bacterial infection that has been rampant in many countries. It is caused by animals which host the infection, typically a rodent, urinating into the ground or in a body of water. Any animal that comes into contact with tainted soil or water is susceptible to receiving the disease. Signs of exposure do not show up right away. It can take roughly one to two weeks before a canine shows symptoms of having leptospirosis, and can stay in the body for several months.

The disease targets the liver and kidneys. The canine will break out in a fever during the first stages of exposure. Symptoms which follow are vomiting, diarrhea, acting lethargic, muscle and joint aches, and the possibility of urinating blood. If the liver is severely infected then the sclera of the canine can turn a distinct yellow. Also in severe cases, the canine will cough up blood and there will be blood in their fecal matter as well.

This bacterial strain is highly contagious to humans. If a person comes into contact with the urine of an infected canine then they are susceptible to contracting the disease. Overall, the infection will lead to liver and kidney failure in both canines and humans. It can be treated in canines with a medication known as doxycycline, and prevented with a leptospirosis vaccination. In the wild, animals are not so lucky in recovering from this infection.

Rabies

This disease is what I would consider to be the scariest outbreak in canines. As most of you may know, rabies is transmitted through being bitten by a carrier of the virus. The most common carriers of the disease are rodents, bats and even foxes. Rabies instantly spreads to the nervous system, but can take close to a month before any of the true symptoms begin to appear.

After a canine has been infected with rabies, the nervous system is heavily impacted upon the virus making its way to the motor and sensory nerves. The disease spreads very quickly and can do one of two things to a canine. It can either begin to paralyze a canine, or it can make them rabid. Sometimes it will do both depending on how it spreads within the brain. The virus essentially eats away at the gray matter within the brain and will ultimately kill the canine it infects.

Some of the symptoms to initially watch out for are

fever, the canine eating things it normally shouldn't or wouldn't consume (a symptom called pica), having seizures, inability to swallow, and becoming hydrophobic. In later stages of exposure the canine will lose its eye focus, the jaw will droop or hang, parts of the canine may become paralyzed when walking, excessive salivating or foaming at the mouth, and shyness or aggression. There is no cure for this virus once it has been transmitted to a canine. It can only be prevented in dogs by staying up to date on rabies vaccinations. Rabies vaccines cannot be given to wolves that are under captivity. It leads to health risks and does not work the same way as it does for dogs. Unfortunately, dogs that contract rabies have to be put down. People who are bitten by rabid canines need to seek immediate medical help. Rabies will affect your brain and nervous system, and can lead to death if exposed to the virus for an extended period of time.

This virus is actually one of the reasons people became afraid of wolves. Many people were encountering packs of wolves that were inflicted with rabies and had turned rabid. This is why you may have read reports of wolves attacking humans throughout history in various countries. Granted, some attacks have been for other reasons, but a rabies outbreak is the leading cause for attacks on humans. Idaho has experienced several instances of wolves that are afflicted with rabies trying to attack humans that are hunting, hiking, farming, etc.

E. Coli

This bacterium strain has spread quickly around the United States within the past decade. E. coli is a bacterium that remains dormant in some animals' bodies, but it becomes active for several reasons. It can be contracted from food or water contamination, being in poor health, which makes the animal more susceptible, coming into contact with infected excrement, or from the animal being unhygienic. The disease will spread to the lower intestine which is where most of the symptoms occur.

A canine that becomes infected with E. coli will start off by being lethargic or unable to move around much. After the disease establishes itself further throughout the intestines it will cause vomiting, severe diarrhea, dehydration, loss of appetite and the body temperature lowering. In severe cases the disease will lead to sepsis.

E. coli can transfer from canine to humans by coming into contact with their feces. If your dog is having these symptoms then it is highly recommended to sanitize the house and wash your hands after touching or cleaning up after the dog. Seeking medical help is also recommended to ensure the dog is not worsening in condition. Puppies and canines with a weakened immune system are severely at risk, as they can die from the symptoms of this bacterial strain. Most canines both in the wild and at home recover from this disease after a couple of months, though.

Salmonella

Many of you have probably heard of this bacterial infection. This disease is typically contracted by a canine eating contaminated or raw food, coming into contact with excrement that's already infected, or being in an inflicted area with salmonella. The bacteria can stay dormant in soil and fecal matter for up to a year. Salmonella has been found worldwide and pandemics of it have spread in recent years. This disease affects the digestive and intestinal tract of both animals and people.

Canines affected with salmonella will have a fever and experience vomiting and chronic diarrhea. Sometimes the diarrhea will be bloody or have a terrible odor to it. This can also lead to dehydration and lethargy. In uncommon cases, salmonella can lead to abscesses that grow on several vital organs which include the liver, kidneys and lungs. If left untreated in the animal, salmonella can cause weakness or potential death from dehydration.

Wolves are fairly prone to getting this bacterium in northern America where salmon populations are higher. Dogs that are fed a raw diet are susceptible to this infection as well since the bacterium easily settles in uncooked meat. Humans also run the risk of exposure if they come into contact with the excrement of dogs that have salmonella. It's highly recommended to wear a facial mask and gloves when handling an animal that has this bacterial infection.

Sarcoptic Mange

This is a scary infection to come across. Many who visualize mange will picture an animal with a hair loss or the animal having the appearance of some morbid creature. In fact, sarcoptic mange is what started the whole scare of the chupacabra myth. These were merely dogs or coyotes that had the infection and would kill farmers' livestock out of necessity. Rumors spread upon their sightings and the myth of a cryptozoonotic creature lurking around Mexico and the southern United States developed.

This infection is actually caused by a mite known as

scabies which burrows into the skin and eats away at the animal. Symptoms to this skin disease are easily distinguishable. The canine will have hair loss all over its body due to consistent scratching from skin irritation. The skin will look wrinkled and even crust over around infected areas. Rashes can also appear all over the body of the animal as well. This infection isn't lethal to the canine, but it will continuously cause irritation and rashes to develop.

These mites will spread from canine to canine and is considered highly contagious. Humans exposed to animals with mange are highly susceptible to having the mites transfer onto them. While people do not suffer hair loss from scabies, it will cause severe skin irritation. I dealt with a dog that was infected with scabies and it is not a situation you want to expose yourself to. Always wash your hands and any exposed body parts thoroughly after touching an animal with mange. There are a ton of treatments to deal with scabies so ask your veterinarian for their recommendation if an outbreak occurs, and your doctor if you become exposed to it.

Tularemia

This bacterial strain is sometimes known as rabbit fever.

The way this disease is contracted is by ingesting a smaller animal that may carry it, or being bitten by an insect that carries the infection. Ticks seem to be the leading carrier to this bacterium. A canine can also get this disease by drinking stagnant water, exposure to contaminated soil and even inhalation. It's mostly seen in dogs but it can be transmitted to any animal, as well as humans. Tularemia is found worldwide and has been popping up more recently in various countries.

The side effects of this infection set in rather quickly. The canine will break out in a fever, become dehydrated, lethargic and lose interest in food. Longer exposure to the disease will cause the liver and spleen to become enlarged, ulcers to develop on the tongue and in the mouth, and the sclera becoming yellow. Canines will become severely ill after long exposure to this infection, with a chance of it becoming fatal if preventive measures are not sought.

Coming into contact with an infected animal can cause the infection to spread from one host to another. Many animals will become afflicted in this way, and humans are no exception. Inhaling the bacteria from an infected animal is the most common pathway to infection for humans, so keep yourself well protected if handling a canine with this disease. A veterinarian will generally prescribe antibiotics to cull the infection from an animal.

Sporotrichosis

This is a fungal infection that is caused by coming into contact with spores found in soil patches. The fungus can only enter a canine's body through cuts or open wounds. Once inside the skin, it will spread throughout the lymph nodes and can even move to the lungs and liver. Sporotrichosis is found in the Midwest and northern parts of the United States. Many wolves, coyotes and hunting dogs are susceptible to this infection.

When these spores first enter the canine's body it will create sores or lesions around the skin which can ooze or rupture. Typically the fungus will remain in the skin and lymph nodes causing severe dermatitis to the inflicted areas. Although rare, the spores have been known to enter vital organs. The liver and lungs are usually the first organs to be affected. This will cause the canine to have a fever, lose weight and have overall lethargy. There is a chance that the fungus can move to the bones, joints, nervous system and even the brain. If you own a dog that comes into contact with these spores and develops sores or lesions then it is highly recommended to seek medical help. There is also a pulmonary form of the spore which can be contracted through inhalation, though it solely results in pneumonia.

Sporotrichosis can be transferred to humans by touching the infected sores and lesions. If you come into contact with a canine which may have this infection then it is strongly advised to wear gloves to avoid any zoonotic outbreak. While this fungal infection is typically non-lethal, there is still the risk of it spreading to the organs within the canine. Treatment will generally last a month to remove the spores from the body, and many veterinarians will use an antifungal saturated solution to remove these foreign bodies from the animal.

While there are certainly more diseases for both wolves and dogs than this list entailed, I felt that the outbreaks provided are the most common and interesting that are seen. Some of these viruses are spreading at an alarming rate and are coming into communities and neighborhoods, affecting our own animals. Infections that we used to see very rarely, such as distemper, mange and even leptospirosis are occurring at a higher rate than dog owners were probably prepared for. This is why it is crucial to get your dog their yearly vaccinations and checkups.

The wolf has to fend for itself in the manner of disease and infection, but does so through natural tendencies. Wolves have a very keen sense of foreboding illnesses and will completely abandon a site they inhibit to keep their pack alive. They are able to smell various outbreaks of bacteria and parasites, and will evade territories where these diseases are present. Not every disease can be prevented, but wolves have the capability of comprehending situations much more judiciously than many other animals. This in itself is reason that I believe wolves have survived for so long and continue to do so to this day.

9

Forensic Science

C rime scene processing is not limited to humans. As threatened and endangered animals are now becoming poached, killed out of revenge for livestock deaths, and trafficked for their body parts, it has increased the need to protect these species. Since wolves are a predator species, they are highly susceptible to this tragic demonstration of human intervention.

Luckily, universities and state parks have opened up positions to help protect wildlife and give justice to the senseless taking of animals' lives. Many of these schools and jobs are trying to reshape what has been lost over the years from environmental destruction, pollution, outbreaks of diseases and the overall detrimental interference with nature. This is what my degree and purpose in graduate school was for; wildlife forensics and conservation.

Forensics plays an important role in specifically analyzing how a wolf, or any animal, died in the wild. Just like crime scene processing that is done for human homicide, a similar, if not the same analyses are done for wildlife. We, as scientists in this field, are able to determine the cause of death for an animal; whether it was killed by humans, some sort of outbreak/pandemic, or even from another animal. I'll now get into how all of this comes into play with protecting wolves around North America.

Whenever a wolf is killed in a protected area or park, there are a couple of things we initially have to look for. Before conducting any sort of forensic processing upon the finding of a deceased animal, tape or paint is set up around the perimeter to ensure nobody disrupts the scene. The range of these perimeters varies on what all is found around the animal, but the standard is a 10 by 10 or 20 by 20 foot range (3 by 3 or 6 by 6 meters).

After this boundary is set up, photography is taken of the body. This is to ensure that all documentation has been taken care of before investigation is done on the remains, and to provide credibility prior to collecting evidence. Courts will sometimes ask for photographs if cases are to escalate. Photos are taken from many angles of the scene, along with any potential evidence around the scene as well. There's a certain method in taking crime scene photography. Various perspectives are required and have to be taken, and there's a right and wrong way to do it. Wide angle shots and medium range shots are done first, followed by close-ups and multiple angle shots around the body.

Once all crime scene perimeters and photos are taken, the next step is to look for impression evidence. This could potentially be one of the most crucial pieces of evidence for any crime scene, and used in lieu of physical evidence that may not be present at the scene. Impressions can range from foot or hand prints, fingerprints, animal prints, tire tracks, indentions of tools or objects that were used and so forth. Photos are taken of anything that can be remotely close to impression evidence. This is so we can hopefully trace back its origin and see if the impression was related to the death or taking of the animal.

Not all crime scenes for wildlife forensics will go this in depth with animal death and homicide. Though, I personally like to be as thorough as possible even if the park or state does not require it. For the most part, wildlife forensics stops all together if it was a single, non-protected animal that has been killed. We are only allowed to conduct further analyses if the state or park seeks retribution for the animals' death. If the park pursues the case, then park rangers may ask visitors if they saw any suspicious activity, if they heard gunshots during their stay, or when they last saw the animal alive.

More often, hunters will kill wolves for game or for trophies. This is one of the biggest contributors to wolf deaths in America. Wyoming recently allowed this type of hunting everywhere in the state aside from Yellowstone National Park, with a limit on how many wolves can be killed annually. Idaho even allows wolf hunting outside of the Sawtooth Mountain range to a degree. Other regions of the country are much more intolerant of wolf deaths, especially those involving Mexican gray wolves and red wolves.

Tagged wolves are also protected by science organizations. Tagging helps scientists track migration patterns and locate established wolf territories. Administering justice for the death of a wolf can be difficult if the species was not protected or if the kill was not done on a reservation or park. Many states will just slap a fine onto the person who killed the animal, but they could potentially face felony charges if multiple wolves were killed or trafficked. It all depends on the state and the court's decision if they wish to pursue the culprit, as wildlife is considered property of the state.

In the states that do wish to pursue criminal punishment, impression evidence can be turned over to police or criminal justice agencies to discover more about whom and what was involved. Footprints generalize the vicinity of where the person came from along with their height, and if they used a vehicle to escape or hull off the animal. This type of detection is considered tracking, which is also used a lot for animal prints.

The lack of an animal's body can sometimes lead all evidence to a dead end. We would only be able to take blood or sediment samples from the scene and run DNA tests if needed. Though, if the animal's body is left behind then there is much more to work with in regards to forensic analysis. Before any samples are collected, bloodstain pattern analysis can be used to determine the time of death for the animal, and potentially what took place at the scene. Fingerprints can sometimes be found on the animal's body, too, but are typically found on firearm cartridges and shells that the hunter may have used. Specific lights, which are either a florescent blue or green, are used to amplify fingerprints. This allows us to take photographs of the prints and send them off to police labs. Prints are only able to be matched up if the culprit has had a previous criminal record. Not every person's fingerprints are in databases, regardless of what you may have seen on television. The person has to already be in the system; otherwise, they become much more difficult to track down through law enforcement.

Sometimes, though, hunters will take the animal to a taxidermist for mounting. At this point, federal law enforcement is able to find out exactly who brought the animal in, seize the evidence, and question the suspect. The remains are then transferred over to wildlife forensic labs, where we are able to verify if the animal belonged to the park or reservation. If proven guilty, fines or an arrest are then able to be administered to the culprit.

Death by firearms tends to be the most reported incident of wolf deaths whenever we come across their bodies. Not a lot of wolves are left behind by hunters who kill them for trophies. Poachers sometimes take the entire animal for their pelts, too. Their bodies are typically left behind by people who commit revenge killings if their livestock was targeted by wolves, or by people who get joy out of hunting predator species for game.

The body is examined for entrance and exit wounds, and if the wolf was shot only once or multiple times. The location of the gunshot is also important in order to see if the wolf was trying to escape or if it was shot out of self defense. These cases are fairly easy to process as the caliber of the bullet, brand of the bullet and its shell/casing tells us all we need to know. It generally shows us what type of gun has been used and whether or not those guns are legally permitted to be owned by the public. If they are not legal to own, then this is when processing gets turned over to local police departments. Otherwise, unless the bullet can be traced back to the owner of the gun, then there really isn't a whole lot that can be done through criminal justice. The only time it's considered an open investigation is if it's a repeated offense with multiple animals killed by the same weapon. A practice known as shooting reconstruction is done to figure out where and how a wolf was killed if we are issued to pursue the case.

Animal impressions are the most fun to analyze in contrast to a human's. Depending on the tracks left behind, it could mean one of two things. Firstly, hunters could have used dogs to track down a wolf for the hunter to kill. Dog and wolf tracks can be hard to differentiate between if you don't know what you're looking at. A dog's prints will sometimes be fairly smaller than a wolf's paw prints. But that isn't the most distinguishing factor between the two. Dogs move their feet side to side, making their tracks run horizontally whenever they walk. Wolves walk with one foot in front of the other, kind of like a catwalk. This makes their paw prints align vertically rather than at a horizontal slant. While hunting, wolves will also place their hind paws in the prints made by their front paws to mask their tracks. This is the best way to differentiate a wolf's tracks from a dog's or coyote's.

Example of wolf prints

If animal prints that do not belong to a dog are present on the crime scene, then it means that another animal was the likely the culprit of the wolf's death. Often times, two wolves from separate packs will fight one another when territory has been invaded, which almost always results in death. In more uncommon occurrences, wolves will go up against different predator animals all together such as bears, mountain lions, coyotes, and even alligators in certain regions of the country. Impression prints and unearthed dirt are observed at the scene along with markings on the wolf's body to tell what it was killed by. In these cases, it's more of an interest to see how the wolf died rather than a necessity for forensic processing. But these fatal skirmishes also help with conservation efforts by mapping out which animals have an allocated territory in specific areas or regions.

If wolves do not have any lesions such as bite marks, gashes, eviscerations, or firearm wounds then forensic processing can become a little more detail oriented. No significant findings on the exterior of the wolf mean that it could have died from disease, parasites, toxins or environmental effects. Whenever this happens, samples from the specimen are taken and sent off to laboratories. These samples can include blood, excrement, urine, organ tissue and even bone fragments.

The majority of these cases involve the wolf dying

from a disease rather than the latter. Depending on the disease, the collected samples are tested to see whether or not the disease is zoonotic, and if it can affect local human populations, or if it's confined to animals in a certain location. If a wolf is suspected to have been inflicted with rabies, its head must be cut off and a study of Negri bodies is done on the brain. These tests are typically done at independent labs such as the CDC. Rabies is the most concerning of any zoonotic disease as it cannot be cured. If there is such an outbreak, local wolf populations must be monitored closely for signs of rabies or rabid behavior. Pandemics are also closely monitored to make sure that there aren't any outbreaks of parvovirus among wolf pups in particular areas. Parvo has occurred several times in the Great Plains and Alaska and wiped out a great deal of wolves.

Parasitic death is not nearly as common with wolves as it is with coyotes and dogs. Although the consumption of raw meat invites bacteria and parasites in, wolves have an exceptional sense of smell and can usually detect numerous and distinctive parasites within food. Their olfactory receptors in the nose are nearly 100 times stronger than a human's. In fact, wolves are able to smell the presence of a parasitic nematode known as *anisakis* in wild salmon. This parasite is known to cause anaphylaxis and severe gastrointestinal pain in both wolves and humans. While wolves most likely do not know what the parasite is per se, they do know it smells 'wrong' and will avoid consuming tainted fish. Many species of parasites can be found in fecal samples of the wolf, which can then be observed under compound microscopes.

Toxicology is used to determine if a wolf has been poisoned or envenomed by something in the wild. There are trappers who try to set poison bait, usually strychnine, out for wolves. They do this to remove wolves from farmlands and towns, but many wolves are aware of these sorts of traps. Snares are the traps that wolves fall victim to the most. They are unable to remove bear traps off their legs before dying from blood loss, infection or exhaustion. The use of poison bait has become either illegal in many areas or is considered controversial. Domesticated dogs have a tendency to consume these poisons around towns, which causes massive uproar from citizens and results in the ban of poison bait.

There is also the concern of poisoning/envenoming from wildlife and flora. However, wolves have become restricted to the northern regions of North America over the years. Therefore, there are fewer fauna and flora species located in these regions that can harm wolves in this way. Even still, canines have a tendency to recover from a considerable amount of toxicity rather than falling victim to any sort of poisoning. Dogs are known to fully recover from venomous snakebites, sometimes with very little harm coming to them.

Serology is used if a canine has been afflicted with a virus, bacteria, or some form of pathogen. This would be where blood or urine is collected to find out what has infected the animal. Hundreds of veterinary and zoological laboratories use these samples to determine internal sickness or for anything involving epidemiology. While some of the outbreaks mentioned in the previous chapter result in death, wolves have a much higher resistance to diseases compared to dogs since they are wild and develop antibodies at a faster rate.

Environmental and natural causes of death are the final studies done. There are times when wolf packs or even lone wolves cannot find an ample amount of food during the winter which causes them to perish. Winter is a very harsh time for all animal species, as food supplies run short and competition between other apex predators increases. The best option wolves have during this season is to work together consistently and separate the weakest prey members from their herds. This is actually beneficial for prey animals as well. The weak and sick animals are the ones killed off by wolves first. If babies are killed then it gives prey animals a chance to breed stronger offspring with hardier mates during the next season.

Aside from starvation, wolves do die from old age, albeit rarely in nature. The average lifespan of a wolf in the wild is between 3 to 5 years, with some alpha members being able to reach the age of 8. Wolves live a much shorter life in the wild compared to captivity simply due to the fact that there are more risk factors found in nature. Signs to look for in these circumstances are hindrances in the musculoskeletal system; weakened or torn muscle tissue, dislocated joints and eroding bone mass. These are the main culprits of wolf deaths around the geriatric age.

Coming across a wolf that has either been killed or has passed away in some form is disheartening to see. It never gets easier, and I simply have to stay focused on what can be done to prevent further deaths. As more people become aware that animals are not something to be taken for personal gain, then perhaps one day we may see wolves reclaim a sliver of their historical habitual land. Several conservation groups that I personally work with are trying to restore their numbers back in the wild. Hope is out there to prevent the death of these animals, but it is up to every single person to become active and help make a difference.

10

Wolf Sanctuaries

The modern gray wolf can no longer be found in some of the areas it used to inhibit within the United States. Species that used to be seen throughout the southern and Midwest regions are no longer found in the wild. Conservation efforts kicked in a little too late for a few of the subspecies of gray and red wolves, which are now facing extinction.

However, optimism is still there for these species and it's all thanks to wolf sanctuaries and wildlife reservations. For those who are unfamiliar with what a sanctuary is, think of it as a protected area for wolves or any other species of animal. They are typically nonprofit groups which aim to restore taxonomic populations and educate the public about conservation efforts. Many of these sanctuaries rely solely on donations and public attendance to pay for the wolf ambassadors on their land. Several of these wolf sanctuaries have around the clock care for their animals and even give housing to their employees to stay on the grounds. Volunteers are always sought to work on the reservation, which is a good way to encourage those wishing to pursue a career in animal science or conservation.

I have worked with a few of these sanctuaries, visited several others and have spoken to various sanctuary directors around the country to help promote and continue conservation efforts with wolves. The employees are all very hands on and work around the clock to ensure the best life for these canines as possible. From a firsthand experience, I can tell that nearly all of these nonprofit groups are very passionate about what they do and seek the best care for their animals. There are numerous sanctuaries that also host foxes, coyotes, wolfdogs and even wild dogs on their property as well.

There seems to be a little confusion in regards to where these organizations get their wolves from. Wolf sanctuaries do not go out and capture wolves to keep them on their property, unlike how various zoos get their animals. A lot of these canine species have either been found or surrendered from people who tried to raise them as pets, or are relinquished from zoos and reservations. It is not uncommon to see many sanctuaries host wolfdogs as they became widely popular at a time for breeding.

As it was brushed on in an earlier chapter, people go out and purchase wolves and wolfdogs thinking they are these tame and exotic animals to own. Shortly after they take the animal home, the animals become destructive, ill tempered or even hostile towards humans. Eventually it reaches a point of uncontrollability for the owner and they surrender the canine to a sanctuary, as no other organizations will take them in. If they are not surrendered or if there are no reservations close by then they are euthanized. Some states ban the ownership of predator species or hybrids thereof. Therefore, if somebody is caught with a wolf, wolfdog, coyote or anything else of similar origin then the state or county could legally seize the animal and put it down.

To help wolves cope with boredom in captivity, they are given enrichment by the staff. This could be anything such as a toy, climbing equipment, uncooked bones to chew on, and even objects with a scent to them. Wolves love to cover themselves in strong smells. In the wild they try to mask their natural smell with anything that has a strong aroma to it. This is to prevent their prey from picking up on their scent and running off. They are especially drawn to the scent if it's unique and if it's something they have never experienced before. One of the wolf sanctuaries I visited used paint and shaving cream for this type of enrichment. Every single wolf loved rubbing themselves on the shaving cream and some of them even tried to taste it!

Wolfdog playing with shaving cream,
Saint Francis Wolf Sanctuary

Some of the ways that wolf sanctuaries collect monetary help are from various tours, photo shoots with the wolves, souvenirs, and symbolic adoptions that are done annually. These are great ways to give back to the organization and have fun helping with conservation. A few of the sanctuaries will even allow you to play with their animals and take pictures with them. There are strict guidelines to follow in doing such, but you can tell everybody that you know that you got to play with a wolf during your visit.

There are a few wolf sanctuaries that I would like to give an honorable mention to. These organizations have personally worked with me through both research and conservation efforts. I feel that they do a wonderful job housing their animals and interacting with the public about the conservation of canidae species. Each of these reservations has a unique way to interact with both the public and their own wolves.

Colorado Wolf and Wildlife Center, Divide, CO

This is one of the first sanctuaries I ever visited and it was a great impact on my desire to get involved with wolves and all other canids. Located approximately 40 minutes west of Colorado Springs, this is one reservation you shouldn't miss out on visiting if you're in the area. Their animals have large enclosures, are very well taken care of and have staff on duty nearly every hour of the day. Much of their supplies come from donations and various contributors. They do an astonishing job at keeping up with their animals' well being. They also inform everybody on their tours about wolf conservation in the wild and how it has been impacted over the years. The staff conducts weekly wolf and conservation reports through social media which is a great way to engage with the public. Their gift shop has something for just about everyone, and for a fee they even let you go into the enclosures with some of the wolves to have your picture taken with them.

Saint Francis Wolf Sanctuary, Montgomery, TX

Located about an hour north of Houston, this is one of the only wolf sanctuaries found in Texas. They are very receptive to their visitors and will answer any and all questions that you may have about wolves. Even outside of their tours they are very active with the community and love sharing information about wolves through social media. Many of the animals on their sanctuary are wolfdogs that have been surrendered or found, and each canine has a unique background. The staff is very involved with their wolves and wolfdogs, providing enrichment for them on a daily basis. They even have their animals paint on canvases with their paws and their bodies, which are then displayed and sold in the gift shop. They're looking to expand their reservation in the near future to have bigger and better enclosures for all of their animals, too. Overall, this is a fun sanctuary to visit if you're ever in the area.

Wolf Creek Habitat, Brookville, IN

This sanctuary is about an hour away from Cincinnati, or an hour and a half from Indianapolis. Wolf Creek Habitat has been around since 1996 and the owners are some of the most down to earth people I have ever worked with. The majority of their wolves have been raised to socialize around humans, many of which they have been taken care of since pups. As it is nonprofit, the sanctuary operates heavily on donations. There are deer hunters that bring the sanctuary carcasses throughout the year for the wolves to eat, while all other food is donated or purchased from tour fees. Their packs range from 3 to 5 wolves, all of which have established hierarchical structures. For a very small fee, the sanctuary allows you to interact with their wolves and take pictures with them. This is something rather unique as very few sanctuaries allow people to come up close to their wolves. They also have a wolf sponsorship program which helps out immensely. One of the most interesting aspects to the sanctuary is that it is located on sacred grounds, and the owners take great pride in protecting and maintaining the land. If you are ever in the area then this sanctuary is a must see for all ages.

Wild Spirit Wolf Sanctuary, Ramah, NM

In a small town nearly 3 hours away from Albuquerque and next to several Native American reservations lies a remarkable wolf sanctuary. This team's mission is to rescue wolves and wolfdogs that have either been abandoned, surrendered or have an underlying issue where they cannot return to the wild. Not only do they house wolves but they also care for several canine ambassadors, such as dingoes and New Guinea Singing dogs. This sanctuary is very attentive to the public. They go out of their way to conduct assemblies for schools to educate kids about the importance of the canidae species and their conservation. As it is a nonprofit organization, much of their funding comes from tours, merchandise and donations. This wolf reservation dates back to 1991 and has had a great impact on wolf conservation ever since. Not enough can be said about the amount of work that this sanctuary does and the dedication the staff and volunteers put into caring for their animals.

Wolf Mountain Sanctuary, Lucerne Valley, CA

This phenomenal wolf sanctuary is located in the high desert of Southern California, nearly an hour and a half to the northeast of San Bernardino. It was founded in 1986 and has roughly 11 wolves on the reserve that are currently being cared for. The staff allows visitors to go into the enclosures with the wolves where they can be interacted with. The sanctuary does this not only for the experience, but to also teach visitors that wolves are animals to be appreciated and respected. Wolf Mountain Sanctuary solely relies on volunteers to take care of the wolves, maintain the upkeep, provide tours, and run the website along with anything else that may be required on a daily basis. As it is a nonprofit organization, every penny from donations goes directly towards the wolves and the reservation they're on. They host several donations throughout the year to help with funding and even have a membership program that you can join. This is an amazing sanctuary with a great history, and the staff is very passionate about their wolves.

There are so many other remarkable sanctuaries that I have not included, visited or worked with yet in North America. Every sanctuary and reservation has a unique background, and is home to wolves with a lot of personality. These organizations are some of the best ways to see actual wolves without having to adventure into the wild for a glimpse at them, along with contributing to wildlife conservation.

Outside of sanctuaries, there are many other ways to contribute to the preservation of wolves. Dozens of reputable nonprofit organizations exist to ensure that all money goes to the animals and not the pockets of employees or governmental bodies. Money isn't everything, though. Staying active with conservation is the most vital resource we have to protect and secure the longevity of all animals that we care about. Bring attention to how animals are treated by county, state, or even at a federal level. Every step in protecting a species is the path to a better future for all of us.

The wolf is said to be the greatest symbol of the wild. They are a representation of absolute freedom and solidarity throughout the world. I urge every one of you to protect this attribute. Keep younger generations' minds open and let them see what the world has to offer. Teach them the importance that every person, and every animal's life matters. Only then may we transcend into something more; a symbol of freedom.

About the Author

Michael Phife is an accredited wildlife scientist and primarily deals with vulnerable and endangered species. His studies were in biology, zoology, wildlife forensics, conservation, environmental science, and marine biology throughout his Master of Science degree. He currently resides in the United States and travels to the mountains, coastal plains, islands and national parks for research. Michael has worked with diverse wildlife throughout his career, ranging from the wolves of North America to free diving with sharks in the Pacific Ocean.

Index

A

adapt
   adapting...15, 23, 40, 89
adaptation2, 18, 23, 42, 58, 68
adaptations
   adapted..........................2
Africa....................1, 5, 71
African wild dog.....71, 74
Alaska
   Alaskan9, 47, 49, 51, 52, 81, 121
alpha
   alphas11, 13, 39, 74, 123
anatomy 3
ancestor
   ancestors 4, 9, 19, 71
anisakis 121
anxiety................31, 32, 38
Asia 2, 4, 19, 31, 73
Australia...............1, 68, 69

B

bacteria..89, 104, 105, 109, 111, 121, 122
bacterial.....98, 99, 101, 102, 105, 106, 108
bacterium...104, 106, 108

Beringia
   Beringian.......................9
Beringian gray wolf.........9
biologists 32, 45, 54, 57
Biologists.......................44
breeding
   breed...9, 12, 17, 18, 19, 21, 24, 25, 26, 28, 56, 61, 76, 127

C

Canada...42, 44, 47, 49, 52, 63, 97
canid...............6, 55, 63, 71
canidae1, 2, 4, 56, 90, 129, 132
canids...5, 9, 10, 57, 89, 129
canine1, 2, 3, 4, 6, 7, 10, 14, 15, 16, 17, 20, 21, 64, 74, 75, 84, 90, 91, 93, 94, 95, 96, 97, 98, 99, 100, 101, 102, 103, 104, 105, 107, 108, 109, 110, 122, 126, 127, 130, 132

140

canines.1, 2, 3, 4, 5, 10, 13, 15, 16, 18, 23, 57, 58, 59, 62, 63, 64, 67, 73, 76, 78, 89, 90, 92, 94, 95, 96, 98, 100, 102, 103, 105, 122, 126

cardiovascular..........94, 97

carnivores
  carnivore..................2, 4
  carnivorous..............2, 3, 9

CDC
  Center of Disease Control..........95, 120

companionship
  companion...14, 16, 17, 28, 70

conservation52, 54, 55, 57, 76, 85, 112, 120, 124, 125, 126, 129, 130, 132, 134, 136

Conservation...55, 64, 125

contagious.91, 92, 94, 102, 107

coydogs...........................64

Coydogs..........................64

coywolves..........62, 63, 80

Coywolves...............62, 63

D

deer.14, 43, 44, 52, 74, 131

descend
  descendant, descendants.5, 20, 73

development

developmental10, 16, 26, 34, 41, 76, 89

DHA...............................34

dhole..............2, 73, 74, 75

dholes........................74, 75

Dholes...............73, 74, 75

dietary
  diet......................22, 23

dingo.................68, 69, 72

dingoes...2, 68, 69, 70, 132

Dingoes.............68, 69, 70

dire wolf.................7, 8, 10

disease...90, 91, 93, 94, 95, 96, 98, 99, 100, 101, 102, 104, 105, 107, 108, 109, 111, 120

diseases...89, 90, 100, 110, 111, 112, 123

distemper..............90, 111

Distemper...............91, 92

diverged
  diverge......................56

DNA.5, 15, 17, 18, 19, 20, 56, 61, 63, 71, 80, 116

docile...24, 32, 33, 60, 61, 65, 87

domesticated
  domesticate...17, 18, 21, 22, 24, 25, 32, 35, 59, 67, 68, 69, 70, 73, 75, 81, 83, 86, 88, 93

domestication...10, 15, 16, 19, 21, 22, 24, 36, 40, 60, 65, 75, 76, 82, 89
dominant
   dominance11, 12, 61, 71, 72

E

E. coli 104, 105
Edward's Wolf................7
ehrlichiosis 95
elk.............................43, 44
endangered........47, 53, 54, 55, 57, 64, 73, 75, 112, 136
Endangered Species Act 47, 52, 64
endangerment..........58, 85
envenomed 122
epidemics
   epidemic.....................13
epidemiology 123
Ethiopian wolves 71
Eurasia 1, 5, 9, 15
Europe...2, 4, 19, 31, 73, 85
evidence......18, 19, 113, 114, 115, 116
evolution.........................58
evolutionary
   evolution.......2, 5, 9, 20
evolve.............................89
evolved....................19, 42
evolving.................. 6
extinct..................7, 20, 54

extinction...8, 11, 54, 55, 57, 84, 85, 125

F

fauna......................77, 122
fear.........31, 32, 35, 39, 55
feral 2, 64, 67, 70
firearm..................116, 120
firearms 48, 117
forensic......113, 116, 120, 136
forensics 112, 114, 136

G

gene 61
genes
   gene 23
genetic...2, 15, 18, 20, 22, 25, 65
genetics2, 5, 15, 23, 56, 63
Great Plains42, 49, 50, 121

H

habitats 1, 85
hepatitis 92
hierarchical..................131
hierarchy
   hierarchies.....12, 13, 74
hunting3, 8, 15, 27, 37, 43, 49, 51, 52, 53, 62, 64, 71, 72, 78, 79, 85, 87, 104, 109, 115, 117, 118

hybrid...60, 62, 63, 64, 66, 80, 82
hybrids..56, 59, 62, 63, 64, 65, 80, 82, 127

I

Ice Age..6, 8, 9, 10, 13, 15, 16, 17, 73, 85
Idaho..............45, 104, 115
impression...........114, 115
infection92, 93, 95, 96, 101, 102, 105, 106, 107, 108, 109, 110, 111, 122
instinct.............................11
Inuit................9, 10, 24, 64
investigation........113, 117

L

leptospirosis.......90, 101, 102, 111
lineage..........1, 4, 7, 24, 71
lineages
    lineage....................4, 20
Lyme disease.................99

M

Mackenzie Valley...44, 49, 50, 51
mange..........106, 107, 111
Mexican gray wolf....47, 53, 55, 56
Mexican gray wolves...115
Mexican wolf...........53, 54
Mexican Wolf Recovery Program......................54

Mexico...................53, 107
Miacid
    Miacidae.................4, 9
Miacids
    Miacid....................4, 5
Midwest...98, 99, 109, 125
migrated
    migrate, migrating4, 6, 9
muzzles
    muzzle...................2, 69

N

Native American...65, 132
Native Americans...43, 64
Negri bodies.................120
nematode..............96, 121
neurological.......39, 91, 96
nocturnal..................49, 57
nonprofit125, 126, 131, 132, 134
non-profit........................54
North America
    United States, Canada.. 4, 6, 7, 9, 31, 33, 47, 55, 63, 64, 85, 93, 113, 122, 134, 136
North America..............73

O

offspring....12, 57, 65, 123
olfactory receptors......121
omegas
    omega........................12
organization.........129, 132

organizations115, 126, 127, 129, 134
originated
    origin.....2, 4, 19, 23, 68
outbreak94, 95, 96, 97, 102, 104, 108, 110, 113, 121
outbreaks....89, 93, 101, 111, 112, 121, 123

P

pack
    packs....11, 12, 13, 21, 30, 31, 39, 40, 44, 49, 50, 57, 69, 70, 72, 74, 75, 111
packs..8, 12, 42, 50, 57, 67, 71, 74, 75, 85, 104, 119, 123, 131
painted dog..........2, 71, 72
pandemic......................113
pandemics
    pandemic...........13, 105
parasite.............96, 97, 121
parasites.....97, 111, 120, 121
park43, 44, 45, 48, 113, 114, 115
parvovirus..............94, 121
Parvovirus......................93
pathogen......................122
physiology 4
pointing.................26, 36, 38
poison....................70, 122
poisoned......................122
poisoning....................122

poisons 122
predator
    predators...15, 43, 52, 58, 59, 112, 117, 119, 127
predators44, 72, 74, 78, 79, 123
prey......3, 40, 58, 71, 74, 123, 127
protection17, 43, 45, 64, 70, 78, 79
puppies......................30, 55
puppy...........34, 36, 38, 40
pups
    pup30, 31, 32, 33, 34, 40, 44, 61, 74, 75, 84, 121, 131

R

rabid 50, 103, 104, 121
rabies50, 90, 102, 103, 104, 120
raven..........................83, 84
ravens 83, 84
red wolf 55, 56, 58, 80
red wolves55, 56, 57, 64, 115, 125
Red wolves...55, 56, 57, 63
relatives 1, 2, 25, 32, 53
reservation115, 126, 129, 131, 132, 134
reservations....31, 55, 82, 85, 125, 126, 127, 129, 132
Restoration....................58

S

salmonella............105, 106
sanctuaries
   sanctuary....12, 31, 33,
     39, 55, 59, 60, 81,
     82, 85, 86, 125, 126,
     128, 129, 130, 131,
     134
scavengers
   scavenger...............8, 56
Serology........................122
South America.................1
species..1, 2, 3, 4, 5, 6, 7,
   8, 9, 11, 13, 14, 15, 17,
   18, 19, 20, 23, 28, 43,
   47, 50, 52, 54, 55, 56,
   57, 58, 60, 62, 63, 64,
   65, 67, 68, 71, 73, 75,
   76, 77, 78, 79, 80, 82,
   83, 84, 85, 86, 87, 89,
   90, 94, 112, 115, 117,
   121, 122, 123, 125,
   126, 127, 129, 132,
   134, 136
subordinate........11, 12, 21
symbiotic relationship
   symbiotic relationships
     ....14, 32, 83
symbol...........................134
synapse.....................32, 35
synapses
   synapse.......................25

T

tails

tail.....................2, 62, 84
Taimyr wolf.................23, 24
taxonomic 125
teeth2, 3, 8, 18, 19, 30,
   34, 69, 71, 75, 84
temperament
   temperaments18, 21,
     24, 32, 39, 59, 61,
     65, 87
territory............74, 81, 119
Tibetan wolf...................68
tick...............95, 96, 98, 99
ticks............96, 98, 99, 100
Timber.......................49, 50
toxicity 122
Toxicology 121
toxins 120
trafficking 59
trait.................5, 25, 57, 86
traits...............1, 2, 11, 17,
   21, 24, 25, 28, 65, 68
trap 46
traps........................46, 122

U

U.S. Fish and Wildlife
   Service 54
United States
   US 9, 39, 42, 43, 47,
     50, 52, 53, 55, 56,
     63, 80, 97, 99, 104,
     107, 109, 125, 136

V

vaccines..................89, 103

veterinarian89, 95, 97, 99, 108, 109
veterinarians..................110
virus91, 92, 93, 94, 99, 100, 102, 103, 104, 122
viruses...89, 90, 92, 94, 111

W

wild dogs15, 67, 68, 71, 72, 76, 126
William's syndrome...........22
wolfdog......17, 59, 61, 64, 82, 83, 127
wolfdogs17, 18, 20, 60, 65, 82, 126, 127, 130, 132

Woodland.......................50
Woodlands.....................42
Wyoming...44, 45, 47, 115

Y

Yellowstone43, 44, 47, 49, 53, 79, 80, 84, 115
Yellowstone National Park..............43, 80, 115

Z

zoo....................................87
zoonotic...…...90, 100, 110, 120
zoos 12, 126